HOW TO BUILD AND REPAIR CONCRETE, BRICKWORK, AND STONE

JOHN A. CORINCHOCK
ROBERT SCHARFF

MEREDITH® PRESS
NEW YORK

This edition is adapted in part from BUILD AND REPAIR WITH CONCRETE

This edition is adapted by arrangement with Robert Scharff and the Quikrete Companies of Atlanta, GA, from their book *Build and Repair with Concrete*.

Because of differing conditions, tools, and individual woodworking skills, Meredith® Press assumes no responsibility for any injuries suffered, damages or losses incurred during, or as a result of, the construction of these projects.

Before starting on any project, study the plans carefully, read and observe all the safety precautions provided by any tool or equipment manufacturer and follow all accepted safety procedures during construction.

Cover Photo by: Portland Cement Association

Published by Popular Science® Books
150 East 52nd Street
New York, NY 10022

For Meredith® Press:
Director: Elizabeth P. Rice
Assistant: Ruth Weadock
Executive Editor: Connie Schrader
Editorial Assistant: Carolyn Mitchell
Production Manager: Bill Rose
Technical Consultant: Gene Schnaser

Produced by Scharff Associates, Ltd., New Ringgold, PA 17960
Production Manager: Marilyn Strouse-Hauptly
Staff Writers: Dave Caruso and Rick Paquette
Cover Designer and Illustrator: Eric Schreader
Book Designer: Karen Weaver
Typesetter: Ann Markowicz

ISBN: 0-696-11110-1
Library of Congress Card Number: 90-061484

Manufactured in the United States of America

10 9 8 7 6 5 4 3 2 1

CONTENTS

Contents

ACKNOWLEDGMENTS

Special thanks to Mr. Dennis Winchester of the Quikrete Companies, Atlanta, Georgia, for permission to use their book, *Build and Repair with Concrete* as the cornerstone for this expanded and revised version.

The following companies and associations also graciously supplied photographs and materials:

Brick Institute of America, Mr. Charles N. Farley
FLR Paints, Inc.
General Electric Plastics
Glen-Gery Brick
Goldblatt Tools
Harry Wicks Photos
J. Rodney Wyatt
Master Builders Technologies, Inc.
Portland Cement Association, Mr. Bruce D. McIntosh
Schuylkill Stone, Inc.
United Gilsonite Laboratories

WARNING: Freshly mixed cement, mortar, concrete, and grout can cause skin irritation. Do not take internally, and avoid contact with the skin whenever possible; the use of rubber gloves, long-sleeved shirts, protective goggles, etc., is recommended. Wash any exposed skin areas promptly and thoroughly with water. If irritation persists, contact a physician. If cement or any cement mixture gets into the eyes, rinse immediately with fresh water for at least fifteen minutes, then get medical attention. Any inhalation of dust can be harmful to your health; repeated inhalation of dust can cause delayed lung injury (silicosis). Keep children away from cement powder and all freshly mixed cement products. As with any job, standard safety precautions should be taken at all times.

Chapter 1

UNDERSTANDING CEMENT AND CONCRETE

Concrete is an extraordinary building material. Its natural beauty and clean, distinctive lines blend smoothly with most architectural styles and landscape designs. Well-designed and constructed concrete projects, such as patios, driveways, landscape walls, and walkways, add lasting value to your property and enhance its appearance.

As a building material, concrete offers unmatched versatility and almost magical working properties. For example, name another building material that can be molded into any one of a thousand shapes. Or one that can be finished smooth, textured, or imprinted to have the look and feel of brick, tile, slate, or stone. Concrete can be white, tinted the colors of the rainbow, or embedded with decorative stone finishes. It can be cast into pavers, building blocks, and dozens of other useful items.

Can you think of any building materials that are stronger and last longer than concrete? Like stone and masonry work, concrete's lifetime is measured in generations. And since concrete is made of inorganic materials, it cannot be attacked by termites, ants, or rodents. Concrete is noncombustible, and when properly mixed and placed, it is unaffected by extremes in heat and cold. For footings, foundations, and other brute strength applications, concrete cannot be surpassed.

The cost of concrete also compares favorably with other building materials. This, along with its other advantages, often makes concrete the building material of choice among professional architects, contractors, and landscape designers. It is constantly selected for all types of industrial, commercial, and residential building and renovation. But what many homeowners fail to realize is that dozens of practical concrete projects usually left to the professional are well within their capabilities.

In almost all cases, concrete is molded—placed into a prepared form and permitted to harden. Slabs, steps, and walls are nothing more than molded concrete shapes. The form for the mold is usually made of wood, so concrete work involves a bit of carpentry. But in this type of basic carpentry, the need for strength in construction far outweighs the need for absolute precision and fine joinery.

All concrete work involves placing a quality mix into sturdy formwork.

Concrete is made according to a recipe, like the batter for a cake. The recipe may vary slightly, depending on where you live or on the concrete's final use, but once you understand the role of each ingredient and how it interacts and affects the others, batching a good mix is no problem.

Floating and troweling concrete surfaces requires skill and a certain feel for the work. But this skill and precision is not beyond the average homeowner who is willing to pay attention to detail and work from smaller projects to more ambitious goals.

The first step in becoming comfortable with concrete is understanding what it is, how it is made, and how it behaves. Armed with this knowledge, you'll be able to avoid many problems that can occur during concrete work.

A QUICK LOOK BACK

The widespread use of concrete dates back to the turn of the century, but when you place a concrete sidewalk or driveway or construct a concrete foundation or footing, you're actually using building technology that dates back thousands of years. Ancient Romans developed cement and concrete similar to the kinds used to-

Finishing concrete is an acquired skill. Start with smaller projects and work your way to larger jobs.

day. By blending a mixture of water, lime, and volcanic ash called *pazzolana*, they produced a hydraulic cement that hardened under water. It became an indispensable building material for ancient roadways, walls, and aqueducts. Some of these structures still stand today, monuments to concrete's dogged durability.

Unfortunately, people lost the art of making cement after the fall of the Roman Empire. It wasn't until 1756 that John Smeaton, a British engineer, again wrote of working with naturally occurring hydraulic cements. In 1818, Canvass White, an American engineer, discovered rock deposits in central New York state that made natural hydraulic cement with little processing. Large amounts of cement made from this rock were used in building the Erie Canal.

The modern cement industry traces its roots directly to Joseph Aspdin, a British bricklayer. In 1824 Aspdin discovered he could make a cement superior to natural cements by carefully mixing, grinding, burning, and regrinding select amounts of limestone and clay. He called his cement *portland cement* because it produced concrete the color of limestone on the Isle of Portland, a peninsula in the English Channel.

Although Aspdin patented his formula and process, astute entrepreneurs soon altered his formulas slightly and set up production. The first recorded shipment of portland cement to the United States was in 1868, and David Saylor established the first American production plant in Coplay, Pennsylvania, in 1871. By 1898, ninety-one different formulas were being used to manufacture portland cement in this country. In 1917, the U.S. Bureau of Standards and the American Society of Testing Materials established standard formulas for portland cement production.

CEMENT VERSUS CONCRETE

Cement and concrete are not one and the same. Portland cement is only one ingredient in the concrete mix. The others are aggregates, such as sand, gravel, and crushed stone, plus water. The aggregates give strength and volume

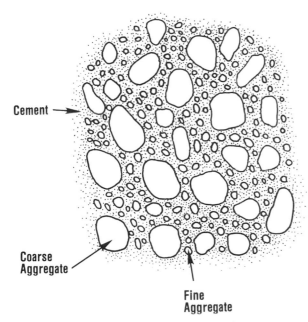

Example of aggregate packing in a good concrete mix

to the mix. The cement is the glue or bonder that envelops and holds the mix together; the water is the catalyst that triggers the chemical reaction in the cement known as hydration.

Understanding Hydration

Concrete does not harden as the water in the mix evaporates. Anyone who doubts this statement should remember that concrete will harden under water. Still, this popular misconception probably ruins more first-time projects than any other. A concrete mix depends on its water content for hydration, the process by which the cement hardens and sets. And it depends on water for a long time, a fact even experienced workers sometimes forget.

Hydration begins as soon as cement comes in contact with water. Each cement particle forms a type of crystalline "growth" on its surface that gradually spreads until it links up with the growth from other cement particles. The growth can also grab onto the surface of the aggregates in the mix. This gradual interlocking of materials results in stiffening, hardening, and strength buildup. Heat is also released as hydration takes place.

The mix begins as a moldable mass, capable of filling forms of any size or shape. But in a relatively short time—twenty to forty minutes—hydration progresses to the point where the mix gains its initial set. At this point, the concrete is stiff enough to be finished smooth. And depending on the fineness of the cement particles, the proportions of the mix, and the surrounding temperature, the concrete can lose all of its workability in as little as an hour.

But this does not mean that hydration has stopped. Hydration continues as long as space exists for hydration products, and temperature and moisture conditions are favorable. Consciously maintaining these ideal conditions by keeping the concrete moist and covered to prevent rapid surface evaporation is known as *curing*. For most work, this formal curing period lasts three to seven days, but the concrete will continue to absorb moisture from the air (provided it's there) for much longer. Most of the hydration and strength development takes place within the first month of the concrete's life. But they continue, although much more slowly, for a long, long time. In fact, laboratory studies have found strength increases over a fifty-year period.

Obviously, misting down your new driveway every evening for the next fifty years is hardly worth the effort. But neglecting curing, particularly in the first critical week, can seriously affect the final quality of the job.

CONCRETE INGREDIENTS

Concrete is basically a mixture of active and inert ingredients. The active ingredients are the portland cement and water. Sand and stone aggregates make up the inactive or inert materials.

Portland Cement

Portland cement is not a brand name. Joseph Aspdin's original name for the particular type of cement remains, and his manufacturing process, although more sophisticated today, is fundamentally the same.

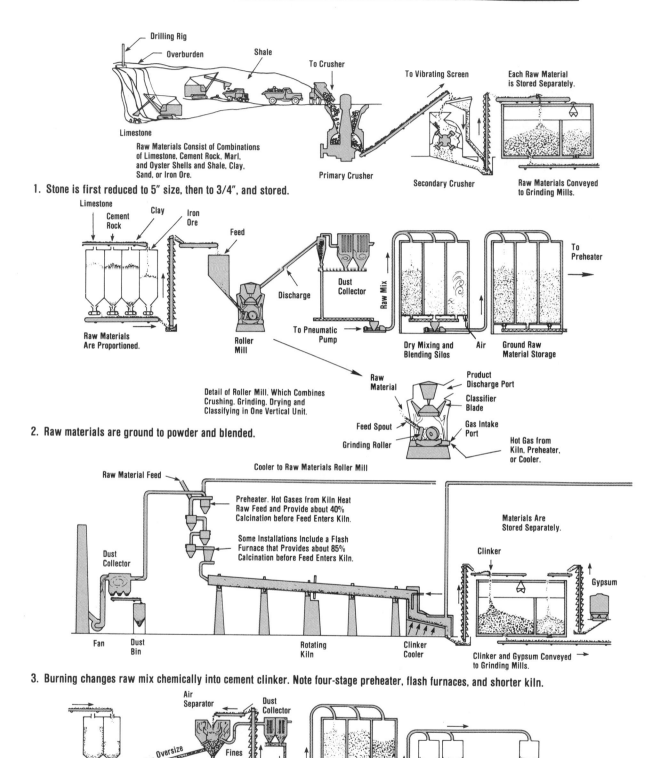

1. Stone is first reduced to 5" size, then to 3/4", and stored.

2. Raw materials are ground to powder and blended.

3. Burning changes raw mix chemically into cement clinker. Note four-stage preheater, flash furnaces, and shorter kiln.

4. Clinker with gypsum is ground into portland cement and shipped.

Steps in dry-process cement manufacture

To make portland cement, selected raw materials are crushed, milled, and proportioned so that the resulting mixture has the desired chemical composition. The raw materials include calcium oxide in the form of limestone, chalk, or shells, and silica and alumina consisting of clay, shale, or blast-furnace slag. The materials are ground and blended using either a dry or wet (slurry) process. After blending, the ground raw material is fed into a kiln and heated to 2600°F to 3000°F. This heating chemically changes the material into cement *clinker*—grayish pellets about the size of marbles. The clinker is cooled and ground to a powder so fine that most of it will pass through a 200 mesh sieve having 40,000 openings per square inch. This fine grind increases the contact surfaces between the cement and water during hydration. Small amounts of gypsum are also added to the pulverized cement to help regulate the setting time.

Five types of portland cement are produced.

Type I: Normal Portland Cement.

Type I is the blend most often used in home and light commercial applications. It's ideal for all general construction work, such as sidewalks, patios, steps, walls, and foundations. Most normal cement is gray, but Type I decorative white cement is available. White portland cement is made from raw materials containing little or no iron or manganese, the substances that give ordinary cement its gray tint. White portland cement is used in stucco, cement paints, finish coat plasters and cement toppings, and decorative concrete. To obtain the truest color possible, white cement should always be used whenever you plan to tint or color the concrete during the mixing process.

Portland cement Types II through V are used when specific problems or conditions exist on the job.

Type II: Modified Portland Cement.

This blend of cement is used for structures placed in water or soil containing moderate amounts of sulfates. These sulfates can generate sulfuric acids, which attack and deteriorate normal blends. Type II cement also offers a moderate reduction in the amount of heat generated during hydration. This is sometimes needed to help control the curing of large piers and heavy abutments and retaining walls. Type II portland cement is seldom used in home applications.

Type III: High Early Strength.

Type III provides high strengths at an early period, usually a week or less. It is chemically and physically similar to Type I except that its particles have been ground even finer. It is used when forms have to be removed as soon as possible or when the structure must be placed into service quickly. In cold weather high early strength cement permits shortening of controlled curing times. Type III is most often used in commercial construction, where time is truly money. For most work around the home, Type I, and a little patience, works just as well.

Type IV: Low Heat Portland Cement.

This is another blend used only in heavy construction trades. It is formulated to minimize the heat of hydration in massive structures, such as dams.

Type V: Sulfate Resistant Portland Cement.

This blend offers maximum protection against sulfate attack.

Air-Entrained Portland Cements

One of the greatest advances in concrete technology was the development of air-entrained concrete in the mid-1930s. Special air-entraining agents can be blended into the cement during manufacturing or added to the mix water. These agents allow extremely tiny air bubbles to form in the concrete as it sets up. The diameter of these air bubbles can be as small as 4/10,000th of an inch. But there are so many of them that they can make up to 6 to 8% of the concrete's total volume. Air-entrainment produces some very desirable characteristics, including:

● Increased resistance to damage caused by freeze–thaw cycles

- Significant reduction in scaling or flaking caused by the application of chemical deicers or salts
- Enhanced workability for easy handling and finishing
- Reduced aggregate segregation and water bleeding in freshly placed concrete

Because of the freeze–thaw and deicer protection air-entrained cements provide, they should always be used in northern and moderate climates. Portland cement Types I, II, and III are available as air-entrained cements. Look for the letter A after the type listing on the bag.

Handling and Storage

If you plan to mix your own concrete, you'll purchase the amount of needed portland cement in 94-pound paper or cloth bags. When freshly packed each bag has a volume of about 1 cubic foot.

Portland cement is moisture sensitive, but if kept dry it retains its quality for long periods. Cement that has absorbed moisture from the air or ground sets more slowly and has less overall strength than cement that has been kept completely dry. The best way to ensure that cement is dry is to buy it when you need it and avoid storage problems altogether. If you must store cement for extended periods, store it indoors on a raised pallet or on planks. Keep the bags snug against one another to reduce air circulation around them, but never stack bags against an outside wall. A tarp or other waterproof covering offers additional protection.

Cement stored for long periods can develop what is called warehouse pack. You might encounter this in a bag purchased from a dealer with a large inventory. The problem can usually be corrected by rolling the bags on the floor to break up the compacted cement powder. At the time of use, the cement should be free flowing and free of lumps. If the lumps do not break up easily, or if the bag of cement shows signs of water infiltration, do not use it for important work. For example, cement that has suffered light moisture infiltration is fine for setting fence posts but is too risky to be used for foundation and other load-bearing work.

AGGREGATES

Aggregates serve as the inert filler material in the concrete mix. Consisting of sand, gravel, and crushed stone, aggregates make up 60 to 80% of the concrete's volume. This significantly reduces the amount of cement needed, reducing the overall cost of the mix. Aggregates are divided into two distinct classes: fine and coarse.

Fine Aggregates

Natural sand is the most widely used fine aggregate. Concrete sand has particles ranging in size from 1/4" down to dust-size specks. Mortar sand should never be used to mix concrete because it contains only smaller particles. Keep this distinction in mind if you plan to mix your own concrete from scratch.

Coarse Aggregates

Gravel and crushed stone are the most widely used coarse aggregates. Rounded pieces are

The ideal aggregate contains a full range of particle sizes from very fine to maximum coarse diameters.

better than long, sliver-shaped pieces. In general, gravels are smooth and round in comparison to crushed stone, which is rougher and more angular. Rough-shaped aggregate produces concrete that is a bit more difficult to work and finish than are mixes with rounded stone or gravel. Cutting back a little on the rough-shaped aggregate in the mix will help solve this problem.

Coarse aggregate ranges in size from 1/4" to 1-1/2". The exact size range used depends on the job. Obviously, the most economical mix is obtained by using a range with the largest maximum size practical. This maximum size is usually equal to one-quarter the thickness of the finished concrete. So a 4" slab might use coarse aggregate with a 1" maximum size, and a 6" slab can safely use aggregate up to 1-1/2" in size. Use 1" maximum size aggregate when constructing concrete steps.

Remember, these are maximum size recommendations only. The best concrete aggregate has a full range of sizes from tiny sand particles up to the maximum practical size. The larger pieces fill out the bulk of the concrete, minimizing the use of the more costly cement. Smaller aggregates and sand help fill in the spaces between the larger aggregates and enable the surface to be finished smooth and hard. An even distribution of aggregate sizes produces the strongest, most economical, and workable mix.

Quality

Both fine and coarse aggregates must be clean and free of dirt, clay, silt, coal, or other organic matter such as leaves and roots. Foreign matter prevents the cement from properly binding the aggregates together, resulting in porous concrete with low strength and durability.

All coarse aggregate should be hard and solid with no flaking or crumbling pieces. If you suspect that the sand you are using contains too much extremely fine material, conduct a silt test. Fill an ordinary quart canning jar or similar clear glass container to a depth of 2" with a representative sampling of sand taken from at least five different locations in the sandpile. Add clean water to the sand until the jar is about three-quarters full.

Shake the container vigorously for about a minute, using the last few shakes to level off the sand. Allow the container to stand for an hour. Any clay or silt will settle on top of the sand. If this layer is more than 3/16" thick, the sand should not be used.

Buying and Storage

If you plan to "batch" your own concrete, you'll have to buy and store the aggregates needed. Purchase the fine and coarse aggregates separately, never mixed. If there is a ready-mix concrete company in your area, inquire about purchasing aggregates from it. The aggregates they sell are the same aggregates they use, so the quality and size range are guaranteed to make good concrete. Of course, you can also purchase materials from a reputable building materials dealer.

Store aggregates on a clean, hard surface. Try to avoid placing them directly on the ground. Ground storage can cause contamination with dirt, grass, and moisture. Cover exposed aggregate to keep out rain, and never use the bottom layer of an uncovered aggregate pile.

WATER

Water is the second active ingredient in the hydration process. Water that is drinkable with no pronounced taste or odor is fine for making concrete or mortar.

If you are drawing water from a pond or stream, be sure it does not contain algae. Algae can cause a great loss in strength in the finished concrete by either influencing hydration or causing an excessive amount of air to be entrained in the concrete. Algae can also be present on aggregates that have been improperly stored or handled. Algae-contaminated aggregate does not bond well to the cement paste.

KEYS TO QUALITY CONCRETE

Strong, durable concrete projects result when good quality materials are placed with care and competence in each step of the construction sequence. The success of your project, whether it is an elaborate patio with an exposed aggregate finish or a simple slab for your satellite dish, is affected by several key factors, including:

- Proportions of the concrete ingredients
- Characteristics and quality of the aggregates
- Presence or absence of entrained air in the mix
- Workmanship of the crew placing the concrete
- Proper curing of the concrete

With care and planning, all of these factors are within your control. Let's begin by taking a closer look at the qualities of good concrete. A simple definition of good concrete is concrete that is relatively easy to place, tolerant of some misuse, capable of taking a good finish, strong and hard enough to support any applied loads, and durable under the most severe local exposure conditions.

The quality and strength of the finished concrete depends a great deal on the quality of the water–cement paste. When the cement and water components of the concrete mix are properly proportioned, a number of good things happen automatically.

Water–Cement Ratio

The water–cement ratio is simply the weight of the mix water divided by the weight of the cement in the mix. For example, if your mix contains one 94-pound bag of cement, fine and coarse aggregates, and 47 pounds of mix water, the finished concrete would have a water–cement ratio of 47 ÷ 94 = 0.50.

Concrete is strongest when the amount of water used in the paste is kept to a minimum and the water–cement ratio is as low as practical. A low water–cement ratio will increase the concrete's watertightness and resistance to weathering. The concrete will experience less volume change from wetting and drying and less cracking due to shrinkage. And concrete with a low water–cement ratio bonds better to iron reinforcement and existing concrete surfaces.

Concrete Workability

The ease of placing, consolidating, and finishing freshly mixed concrete is called workability. Unfortunately, concrete with the lowest possible water–cement ratio often mixes up stiff and hard to handle. Experienced professionals often work with these harsher mixes using special vibration tools and their considerable skills during the finishing process.

But less experienced workers and certainly first-time novices need concrete with maximum

Workable concrete is easily finished, as in the edging process shown here. The slight shine on the concrete's surface indicates some bleed water is present.

workability. This translates to a wetter, more flowing mix. So for the sake of workability, most concrete mixes contain more water than they actually need for complete hydration.

The minimum water–cement ratio (by weight) for complete hydration is 0.22 to 0.25, but most concrete used around the home has a water–cement ratio between 0.40 and 0.60.

Too much water in the mix also adversely affects its workability. The concrete should be workable but should not segregate or bleed excessively. Bleeding is the migration of water to the top surface of freshly placed concrete caused by the settlement of the solid materials—cement, sand, and stone—within the mass.

A certain amount of bleed occurs in all fresh concrete, but excessive bleeding increases the water–cement ratio near the top surface and a weak top layer with poor durability can result, particularly if you attempt to finish the concrete with the bleed water present.

You can usually judge fresh concrete's quality and workability by its look and feel. When properly proportioned and mixed, each particle of aggregate will be completely coated with paste and all spaces between aggregates will be filled with paste. Freshly mixed concrete should be a plastic, semifluid mass capable of being molded or shaped like a lump of modeling clay. The stone and sand should not tend to segregate or separate out during transport and handling. The concrete should flow sluggishly without separating or crumbling.

Slump

Slump is a measure of the concrete's consistency. Stiff, harsh mixes have lower slump than wet, easy flowing concretes. An official slump test can be performed by filling a special test cone with a sample of the mix. The cone is hollow on both ends and stands 12" high. The base diameter is 8" and the top diameter is 4".

The wet concrete is loaded into the cone in layers and tightly compacted using a long rod to remove all air pockets. The concrete is leveled

A

B

(A) This concrete falls in the 3" to 5" slump range, used for most home projects. (B) The slump of this mix is too great. It is too wet and runny to produce good results.

with the top of the cone, which is then lifted straight up and away. The distance the concrete sags or slumps is then measured in relation to its original height. For most home projects, concrete with a slump of 3" to 5" will produce the proper balance of strength and workability. Keep this slump range in mind when ordering concrete from a ready-mix company.

If you plan to batch and mix your own concrete, a slump test can help determine a good mix, particularly if you haven't worked with concrete before. You can cut the cone body from light sheet metal and hand rivet it together. A small pail or large can may also be pressed into service. The slump reading might not be as precise, but you'll get a good visual indication of the mix's consistency.

Concrete Strength

As mentioned earlier, strength and workability are trade-offs in the concrete mix. Increasing one decreases the other. That's why care must be taken to limit the water content to just the amount needed to generate a workable mix. Adding more water than that will weaken the finished concrete.

Concrete strength is measured as compressive strength. For example, a ready-mix company may tell you their concrete is 4,000 psi concrete. This means that under controlled conditions, a sample of the concrete that has cured for twenty-eight days can withstand a compressive pressure of 4,000 pounds per square inch before crumbling.

Not all jobs require concrete of equal strength. Depending on where you live, your concrete will be exposed to severe, moderate, or mild weather conditions.

Severe exposure includes climates where deicing chemicals are used to melt snow and ice, or where concrete becomes wet and saturated prior to repeated freezings. Obviously, both situations can occur, even in some interior applications. Concrete subject to severe exposure should be air entrained with high strength and low water permeability.

Building codes and general concrete standards impose a low 0.45 water–cement ratio as the maximum permitted for durable concrete in severe exposures. This translates to 5 gallons of water per 94-pound bag of cement. The average strength in concrete with this low water–cement ratio and 6 to 8% total air entrainment is 4,500 to 4,800 psi at twenty-eight days. This is more than enough compressive strength to carry loads placed on walks, driveways, patios, and steps.

Moderate exposure occurs in climates where there is freezing, but where the concrete is not wet long enough to become saturated with

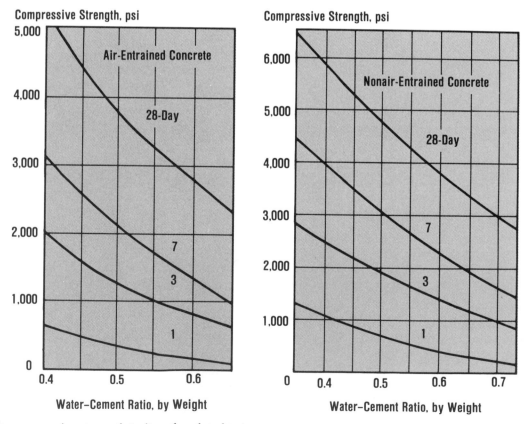

Concrete's compressive strength is directly related to its water–cement ratio. Most concrete used in home projects has a compressive strength between 3,500 and 4,800 psi.

water prior to freezing. Deicing chemicals are not used in moderate exposure conditions. The concrete selected for moderate-exposure locations can be of lesser quality than that required for severe exposure locations, but it still must have low permeability, adequate air entrainment, and moderate strength.

For service in a moderate exposure, a 0.50 water–cement ratio is acceptable, or about 5-1/2 gallons of water per 94-pound bag of cement. When combined with 5 to 7% entrained air, this concrete provides a compressive strength of 4,000 psi at twenty-eight days.

Mild exposure occurs in climates and locations where the concrete is not exposed to freezing or to deicing chemicals. Concrete in mild exposures requires enough strength for structural safety and 2 to 3% air entrainment to give good workability.

Concrete Durability

Good concrete is durable and capable of withstanding seasonal weathering and eroding forces. The top surface of concrete is made up of cement paste and fine aggregates. Lower water–cement ratios will give this top layer a more watertight finish, but no concrete is completely impervious to water penetration unless specially treated.

Damage is caused when water soaks into the concrete and freezes. As it freezes, the water expands, applying pressure in all directions. This pressure can break loose the fine aggregates and cement paste, exposing the coarse aggregates below the surface.

This condition is known as scaling. The situation is compounded by the presence of deicing salts in the water. These salts crystallize during freezing, adding to the internal pressures generated in the concrete.

A number of steps can be taken to avoid scaling problems. The first is to always use air-entrained concrete in areas subject to freeze–thaw cycles and salt contamination. The tiny air bubbles in air-entrained concrete provide empty space for the expanding water. This minimizes the internal forces within the concrete that lead to scaling.

Good workmanship also goes a long way in ensuring durable concrete. Remember, as fresh concrete sets up, some of the water in the mix will always rise to the surface. This bleed water must evaporate before any step in finishing—floating or troweling—is done. Finishing the concrete before it loses its bleed water will work the water back into the top layer and increase the water–cement ratio at the concrete's surface. The end result is the weakest concrete where you need the strongest.

Along with good workmanship, you should also cure the concrete correctly so the top surface will be as strong and durable as the rest of the concrete. And after curing, give the surface about thirty days to air dry before the first winter freeze and application of deicing salts.

CONCRETE'S TENSILE STRENGTH

If concrete has an Achilles' heel, it is tensile strength, or its ability to resist bending stress. Concrete has a low tensile strength (about 650 pounds per square inch). In contrast, ordinary steel has a tensile strength of about 30,000 psi, and some special steels are rated as high as 100,000 psi.

So, unlike a steel girder, which can bend and twist without fracturing, concrete will crack and rupture without proper reinforcement. A major source of reinforcement for all types of concrete work, particularly slabs, is the ground or subgrade. Simply put, the ground or subgrade must be stable and capable of supporting the weight of the concrete and any load placed on it.

Consider a simple slab foundation for a small outdoor storage shed. The weight of the shed and the items stored in it exert a downward force on the slab, pushing it into the ground. If the ground beneath the slab is firm and solid, it can resist this force without movement. Think of the pressure as being transferred through the slab to the ground, with the shed and the weight of the concrete pushing down, and the ground, in effect, pushing up. The slab is compressed between these two forces, but since the compressive strength of good concrete is very high (4,000

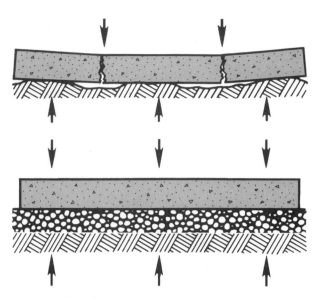

Concrete has low tensile strength or resistance to bending. Good subgrade preparation helps eliminate uneven stress that can crack concrete.

to 5,000 psi), there is little chance of crumbling and failure.

Problems arise when the ground cannot resist these downward pressures equally in all areas. A loose, unstable subgrade simply cannot push back, and differences in the downward and upward forces acting on the slab result in bending action. If the force of this bending action exceeds the tensile strength of the slab, cracking will result.

This is why great care should be taken in selecting and preparing the subgrade for all concrete projects. Never build on fill (recently positioned soil) and avoid areas with considerable groundwater movement, a condition that can shift the soil beneath the work. Details on preparing the subgrade are given in chapter 4.

Steel Reinforcement

Structural cracking can also be significantly reduced by embedding steel bars or steel mesh within the fresh concrete. Steel reinforcement is often used in concrete walls, beams, and larger slabs. The network of steel within the concrete helps absorb bending forces and holds the concrete together when cracking does occur. Steel

reinforcement should never be considered a substitution for proper subgrade preparation.

VOLUME STABILITY

The last property of concrete you should understand is its ability to expand and contract. Hardened concrete changes volume ever so slightly due to changes in temperature, moisture, and stress. Concrete expands slightly as temperature rises and contracts as temperature falls, although it can expand slightly as free water in the concrete freezes. In the same way, concrete expands as it absorbs moisture and shrinks as it dries out. The amount of shrinkage is influenced by the amount of water in the mix, another reason that water content should never exceed the minimum needed for workability.

The magnitude of these volume changes is extremely small, but cannot be ignored. For ex-

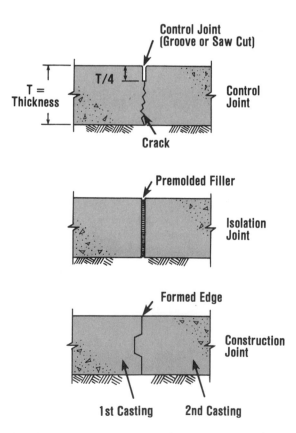

The three types of joints used in concrete work: control, isolation, and expansion

ample, the overall length of a 100′ slab might change 3/4″ due to variations in temperature and moisture content. This volume change is just enough to generate stress forces within the concrete that can crack it.

Cracking due to volume instability cannot be stopped. But it can be controlled by forming or cutting grooves in the surface of the concrete. These control joints force cracking to occur at the bottom of the joint itself, rather than in a random pattern throughout the concrete. In this way, cracks are not visible.

For most smaller projects, control joints are pressed into fresh concrete during the finishing process using a special grooving tool. Control joints can also be cut into the surface of hard-ened concrete using a circular saw and special blade.

Special isolation or expansion joints might also be needed to permit movement between abutting faces of the new concrete and existing structures. And keyed construction joints are the preferred method of tying separate slabs together when larger jobs cannot be completed in one work session.

These joints and the simple mechanics of working with concrete are covered in detail at the beginning of chapter 5. You'll see that most mixing, handling, and finishing steps are designed to complement the unique combination of characteristics just discussed.

Chapter 2

PLANNING, TOOLS, AND SAFETY FOR CONCRETE WORK

Thorough planning is an important part of concrete work. Thinking the job through before you begin, working within your skill level, and paying attention to detail while you work will make the difference between success and failure in both concrete and masonry work.

ESTABLISH LONG-RANGE GOALS

Your home and property should be viewed as a facility that combines leisure, traffic, and utility areas. Each area deserves individual attention. It's a good idea to draw up a master plan for your property. It might take years to complete the individual projects, but in the end all work will be compatible with the master scheme.

Start off by making a scale drawing of the lot and house, regardless of whether the structure is new and not yet landscaped or is established and partially or completely landscaped. Make the drawing a bird's-eye view, like the example shown here. Indicate all boundary lines and structural entries, including garage doors, patio entryways, and the like.

Entryways establish traffic patterns, so walks and paths should be designed to coordinate with them. For example, the outside door adjacent to the laundry room or utility area should exit onto a service walk that leads to an outside clothesline, the trash can area, or free-

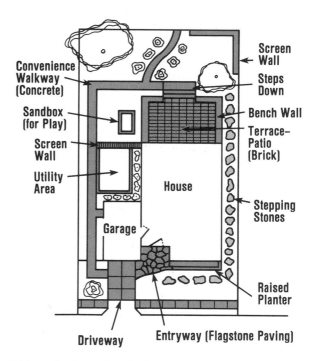

This scale drawing of a house and property indicates numerous concrete, masonry, and stonework projects. Some may be existing; others proposed.

standing tool shed. The width of the walk should accommodate one person who might be carrying a clothes basket or trash container, or pushing a lawn mower or wheelbarrow. A smooth, nonskid surface is ideal.

The entry walk from a driveway or sidewalk to the front door deserves more attention. An eye-catching flagstone, brick paving, or exposed aggregate concrete finish can be used. And the walk should be wide enough to accommodate two people walking side by side.

Garden walks and paths that do not carry service traffic can be built from stepping stones, concrete pavers, or brick. Driveways should be wide enough to handle all vehicles they will serve and allow passengers sure, clean footing.

Once the functional aspects of the master design are identified, you can begin to consider more decorative additions, such as planters, patios, tree wells, and the like. Include all your long-range projects and goals, not just those involving concrete and masonry work.

PROFESSIONAL HELP

Don't be afraid to consult a professional architect or landscape designer for advice about your master plan or for help with larger individual projects. For a reasonable fee, these professionals might be able to point out options in materials and design that you have not considered. They are also familiar with local building conditions and codes. It is their job to visually assess your land and help shape your ideas and

budget into a final project plan that is both functional and visually pleasing.

INDIVIDUAL PROJECT PLANS

Each project should also have its own plan. Ideas for projects and plans are everywhere. (See the color section.) Browse through magazines and books devoted to home and landscape design. Building supply companies also provide brochures highlighting the use of materials they offer.

A fun, easy way to find ideas is to take a leisurely walk through your neighborhood. You'll find it full of examples of concrete and masonry work—some good, some bad.

Once an idea begins to germinate, sketch it on paper. Simple projects may be no more than a set of dimensions jotted down on a scratch pad. A major project, such as a long driveway that gently curves its way up a slope to the garage, or a patio that blends cooking, eating, and lounging areas, will obviously require more formal work drawings than a tree well or planter. Whatever

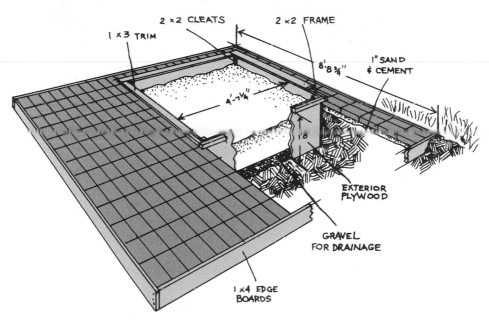

Even relatively simple projects, such as this mortarless brick deck and sandbox combination, should have a formal work drawing to help with dimensioning and material estimates.

the job, take the time to plan it right. A flawed plan will doom the project and waste considerable amounts of time, effort, and money.

Zoning and Building Codes

Most communities have strict zoning regulations and building codes. Zoning ordinances regulate the type and size of building allowed in a specific locale. Zoning laws usually state the minimum distance the building must be set back from the edge of your property. So be sure you understand the local requirements before you lay out the footing or foundation of a garage, outbuilding, or home addition. If you're planning an addition to a dwelling built near a property line before current zoning ordinances were passed, you might have to file for a variance to be excepted from the newer setback requirements.

Building codes govern all aspects of the actual construction. For example, the building code might spell out the dimensions for concrete footings, the quality and strength of the concrete used, and the acceptable methods for construction. Most communities adopt one of the so-called model codes used nationwide, but you should still check for any particular amendments peculiar to your area. Codes can be very specific pertaining to sidewalks and driveways that cross a public way. Some might even set grades and styles.

Never be tempted to ignore codes or modify their requirements as you work. Projects constructed without regard to local codes can be unsafe and illegal. By adhering to them, you'll be well on your way to professional results and will avoid the mistakes and failures that led to the establishment of the codes in the first place.

Building permits are required before you begin construction of driveways, sidewalks, patios, steps, foundations, and almost all other major types of home improvements. Building permits are obtained from local city or county building authorities. Building permits allow local authorities to keep track of construction in the community and ensure that builders are familiar with zoning laws and building code specifica-

tions. Local building departments can also be a valuable source of information. For example, if you are planning to contract out part of the work to professionals, local building authorities can supply a listing of reputable builders in your area. They can also advise you as to the depth of the frost line in your locale and provide additional information on the project you're undertaking. So don't be afraid to ask questions when you go in to pay the permit fee.

Know Your Property

Review the deed to your property. It might include restrictions that can affect what you may build and where you may build it. Finally, pinpoint the locations of all underground utilities on your property, such as gas and water lines, electric and telephone wires, and septic systems. Building on top of these utilities or worse, striking a gas, water, or live electrical line while digging, could be disastrous.

ASSESS THE WORK INVOLVED

Once your project plan has been finalized, take some time to assess the work involved in each step of the job. If the job involves a simple slab to support a small tool shed, a short service walk, or landing at the door to your garage, it's likely you can handle all phases of the work yourself. On such projects, the site preparation should be minimal, the formwork basic, and the amount of concrete to mix, move, and finish undaunting.

The major consideration in concrete work is the amount of concrete you and your crew can handle in a given period. This is particularly true regarding foundations, large retaining walls, extensive slabs, and the like. Discovering halfway through a job that you're in over your head can be very costly in terms of wasted time and materials. And while it's true certain admixtures can be added to fresh concrete to slow hydration, all concrete work is still a race against time.

One way to avoid the chore of handling and finishing a large amount of concrete is to do all work that precedes the pour, and then call in professional help. If you do the layout and design, prepare the subgrade, and construct the forms, you'll realize substantial savings and still enjoy a true sense of creativity and involvement.

This is not to say that well-organized and properly equipped amateur crews cannot handle midsize and larger jobs. They can with great success. But when assembling such a crew, always try to find one or two workers with some experience in floating, final troweling, and edging. They can keep the project moving and serve as advisors to less experienced workers.

WORK UNDER FAVORABLE CONDITIONS

If fresh concrete freezes before it reaches its final set, it will revert to a mound of aggregate. So the best time for doing concrete and masonry work is in late spring and early summer when there is absolutely no chance of frost and freezing conditions. Working in warm weather gives the concrete plenty of time to develop the strength needed to resist freeze–thaw cycles and chemical deicers.

Quality concrete work can be done in cool weather, but low temperatures greatly slow down hydration and retard final setup and strength gain. Using Type III High Early Strength portland cement reduces the critical curing period from seven to three days, but the concrete still must be kept from freezing during this time.

Fortunately, the heat of hydration helps warm the concrete, and the dissolved chemicals in the cement paste lower its freezing point to below that of pure water.

One of the best ways to keep concrete from freezing is to start with a warm mix. Aggregates can be heated and hot water used. Ready-mix that is heated prior to delivery can also be ordered.

Once the concrete is placed and finished, hold in its heat by covering it with insulating blankets or a layer of straw 1' to 2' thick. For indoor work, portable heaters can be used to heat the area around the fresh concrete. If all this seems like a lot of extra work and bother, it is. So save yourself the trouble and plan to work in good weather.

Hot Weather Precautions

Hot weather brings its own set of problems. Precautions must be taken to prevent rapid loss of surface moisture from the concrete. This can cause finishing difficulties and cracking due to rapid shrinkage in the concrete's top layer. To help prevent these problems, do the following:

- Place and finish concrete during the cool early morning or late evening hours. This is also the most comfortable time to work.
- Dampen the subgrade and forms to keep them cool and moist. A dry, hot subgrade will suck moisture out of the fresh concrete.
- Minimize finishing time by having a large crew and adequate tools.
- Avoid working on extremely hot, dry, windy days.
- Use temporary coverings, such as wet burlap or plastic sheeting, to help retain the concrete's moisture. Uncover the concrete in sections as you finish it.
- Mist the surface of the fresh concrete to slow down evaporation, but be careful not to add water to concrete that has not gained its initial set.

TOOLS FOR CONCRETE WORK

Having the right tools at hand and knowing how to use them safely and correctly are important aspects of any job. The tools necessary to do good concrete and masonry work are relatively common and inexpensive. You probably already own some of them, and the rest can be bought, rented, or even made from common household items. However, never underestimate the importance of tools. When you must buy, don't be fooled into choosing cheap, inferior-quality items. Buying good tools and taking care of them will go a long way toward achieving professional-looking results.

Site Preparation Tools

A short-handled, square-faced shovel is ideal for squaring off footer trenches. The long-handled, round-pointed shovel commonly used in gardening is not nearly as effective for concrete and masonry work.

The pickax is useful for loosening sod or soil and for digging footer trenches. Its broad edge makes it easy to remove the top layer of dirt with very little disturbance of the soil underneath.

You'll need a tamper to compact loose soil and firm up the sides of dirt forms, post holes, slab subgrades, etc. Commercial tampers have a heavy steel pad, but you can make wooden ones that will work just fine. A strong metal rake for leveling soil and dumped materials completes the necessary site preparation tools.

Formwork Tools

To make wooden forms, you'll need a good crosscut handsaw or portable circular saw. A

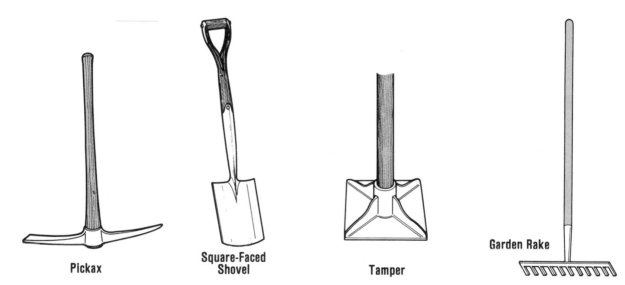

Pickax   **Square-Faced Shovel**   **Tamper**   **Garden Rake**

Site preparation tools

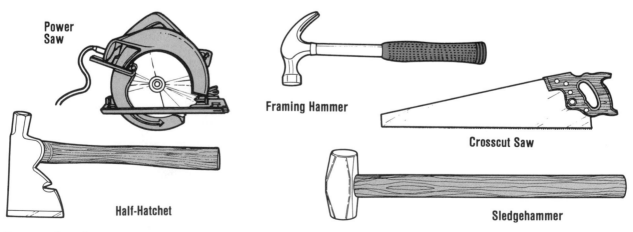

Power Saw

Framing Hammer

Crosscut Saw

Half-Hatchet

Sledgehammer

Formwork tools

20-ounce framing hammer is also recommended, particularly when driving large nails.

To drive stakes, an 8-pound, double-faced sledgehammer is ideal. Choose one with a 24"- to 36"-long handle for heavy-duty work, such as breaking up old concrete or brickwork. And when doing such tasks, be sure to wear safety goggles to protect your eyes from flying chips.

The half-hatchet is a good multipurpose tool to have on hand. The broad, sharp edge is used for chopping points on the ends of stakes, and the hammerhead end is used for driving nails and small stakes.

Layout and Leveling Tools

To lay out the site accurately, set up batter boards, and check forms, walls, etc., for trueness, several tools are needed. The mason's line comes in 200' rolls and is either white or yellow so that it is easy to see. Use the line to mark off perimeters and to outline patterns for forms and grids. A line level hooks onto a stretched line and is used to determine whether two points are level.

Flexible steel tape measures are available in various lengths. It might be a good idea to have two: an 8' or 10' tape for short measuring and a longer one (at least 25') for overall layout and checking. A framing square determines precise 90° angles and can also serve as a straightedge.

A chalk box is used to mark a straight line on a surface. Simply attach the hook to one end, extend the chalk line to the opposite end, and snap the center of the line sharply to transfer the chalk to the surface. The plumb bob is a valuable tool for determining vertical lines quickly and easily. If you plan to do any brick or block projects, a high-quality mason's level is also a good investment.

Mixing Tools

The same shovel used for site preparation is ideal for handling concrete. The square edge is

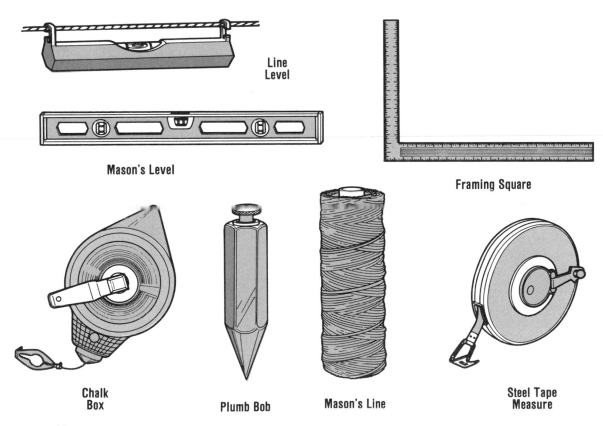

Line Level

Mason's Level

Framing Square

Chalk Box

Plumb Bob

Mason's Line

Steel Tape Measure

Layout and leveling tools

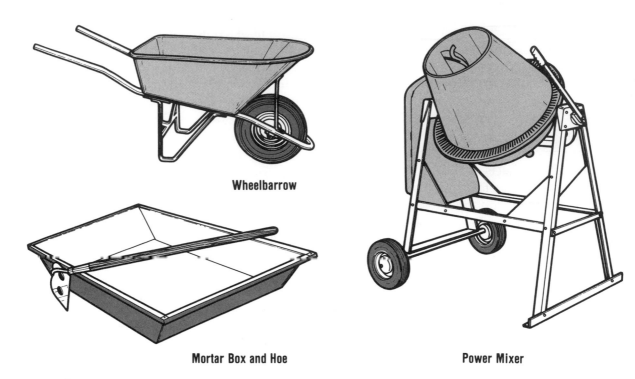

Wheelbarrow

Mortar Box and Hoe

Power Mixer

Mixing tools

useful for working the concrete into corners. A hoe can also be a help; use the mason's type if possible. This hoe features a long handle and a wide blade with two holes in it. The holes make it easier to move the hoe through the ingredients, thus aiding in the mixing process.

A sturdy wheelbarrow can be used for mixing concrete and mortar, as well as moving rubble and other loads around the job site. The mortar box is available in various sizes and is also designed for mixing.

For larger jobs, electric or gasoline-driven power mixers are by far the best way to mix concrete and mortar. Various sizes can be rented to fit the job at hand. To use, simply turn on the mixer, add the concrete or mortar mix to the drum, then slowly add water. Run the mixer for three or four minutes to allow the mixture to reach its proper consistency. Mix no more concrete than you can comfortably place at one time, and mix no more mortar than you can use in a two-hour period.

Concrete Finishing Tools

Screeds, also called strike boards, are used to level the concrete to the height of the forms. They also serve to push the larger pieces of aggregate below the surface. Screeds made of lightweight metal are available commercially, but a straight length of 2×4 lumber works just as well.

When using a screed, move it in a zigzag motion, keeping a small amount of the concrete ahead of the tool to fill in low spots. Tilt the screed in the direction of travel to obtain a cutting edge. If a second pass is needed to remove any remaining bumps or low spots, tilt the screed in the opposite direction.

A wooden or metal hand float and a darby or bull float are used after screeding to give the concrete a uniform surface. Floating uses overlapping arcs to bring excess water to the surface and smooth the concrete. In some cases, floating is the only finishing technique used, resulting in a

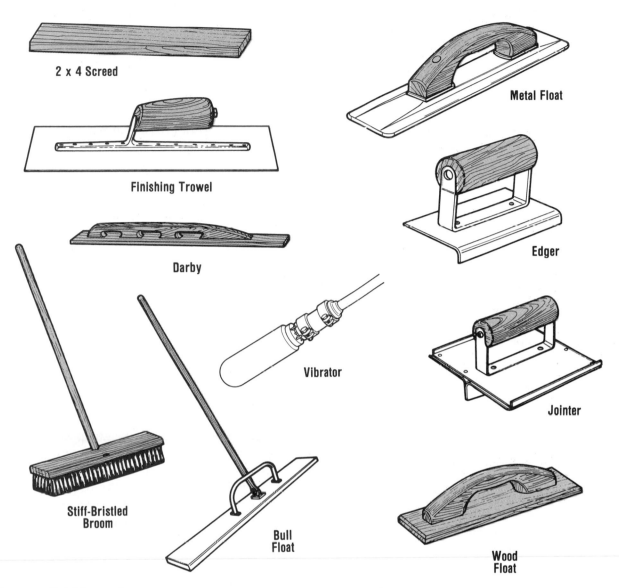

2 x 4 Screed

Finishing Trowel

Darby

Metal Float

Edger

Vibrator

Jointer

Stiff-Bristled
Broom

Bull
Float

Wood
Float

Concrete finishing tools

finish that is reasonably smooth but still has good traction.

A darby is nothing more than an oversized hand float that enables you to cover more area and reach more places without walking on the concrete. Bull floats are larger still and have long handles for working extremely large areas quickly. Many times a darby or bull float is used to prepare the concrete for finishing with a hand float.

Although finishing trowels are designed much like floats, they are always made of high-quality steel. A trowel is used after the float has done its job; its purpose is to provide a slick, hard, dense finish. Never trowel a surface without floating it first, or water will become trapped underneath the surface and cause the concrete to flake.

As stressed throughout this book, timing is very important when troweling. Never begin troweling until all the surface water has evaporated and the concrete has lost its sheen. Hold the trowel at a slight angle and apply even pressure to smooth the surface. Concrete must often be troweled several times to produce the desired finish. For a nonskid surface on floated and troweled concrete, use a stiff-bristled broom.

Edgers are designed to produce a neat, rounded edge on concrete slabs, walks, and steps. The idea is to avoid sharp edges that can easily chip or crack. Edging also hardens the surface next to the form where floats are less effective. The edger is used to cut away the concrete from the form and is then run along it to compact and shape the concrete. A stainless steel edger with a 1/2" radius is a good choice.

Jointers are used to cut control joints in concrete. As mentioned in chapter 1, these joints are needed to prevent cracking as the concrete expands and contracts from the effects of weather and time. To be effective, control joints should extend one-quarter of the way through the depth of the slab. You can also use a portable circular saw to cut control joints.

Internal vibrators are ideal for compacting large areas of concrete, particularly in wall forms. They are especially helpful when the mixture is stiff. Attach the vibrating end to a reinforced hose and lower it into the concrete at 18" intervals along the surface length. Usually ten to fifteen seconds of vibration is enough to compact the concrete and remove any air pockets. The concrete has been worked enough when a thin line of mortar or paste appears near the vibrator.

Tool Care and Maintenance

Wheelbarrows, mortar boxes, and all concrete and masonry tools must be thoroughly cleaned immediately after use. Wash them with water to remove the residue before it has a chance to dry. It's best to rinse your tools several times throughout the workday, then give them a good cleaning at night.

When the tools are dry, apply a thin coat of oil to all metal surfaces to prevent rust. Remove existing rust with steel wool. There's nothing more frustrating than reaching for a tool you haven't used in a long time, only to find it rusted beyond usefulness.

Store your tools on shelves or hang them on hooks, making sure that they are out of the reach of children. Tool boxes that can be locked are also good storage areas. Remember that even the best tools will not perform as they should if they are not cared for properly.

SAFETY

Safety should always be the priority on the job site. Concrete and masonry involves working with heavy, caustic materials. But dressing for safety and exercising a few commonsense precautions will eliminate the potential for personal injury.

Protect Your Back

Sacks of cement can weigh up to 94 pounds. When lifting a sack of cement or premix, your back should be straight and your legs bent. Hold the sack between your legs as close to your body as possible. Do not twist from the waist when lifting or carrying cement sacks or similar loads. Always get help if you need it.

Sand and aggregates are also heavy and clumsy to handle. Take care when loading mixers or pushing wheelbarrows filled with materials. One cubic foot of concrete can weigh as much as 150 pounds, and a cubic yard of concrete—the standard measure for large jobs—weighs in at roughly 2 tons.

A flat-end shovel is the best tool for spreading concrete after it has been placed in the forms. Push the concrete with the blade of the shovel; do not lift and throw it unless absolutely necessary.

Protect Your Eyes and Lungs

Wear proper eye protection when working with cement and concrete to guard against blow-

Wear wraparound eye protection when cutting or breaking up brick, block, stone, or concrete.

ing dust, concrete that might splatter, or other foreign objects. Always wear goggles or safety glasses with side shields when breaking up old concrete, cutting brick or block, or when operating any power tool.

Inhaling cement dust can irritate your throat and lungs, and repeated inhalation can cause delayed lung injury. So wear a respirator mask if you'll be exposed to dusty conditions during prolonged mixing sessions.

Head protection, such as a hard hat, is normally not needed for typical jobs around the home. The one exception is brick or block masonry work that involves working on scaffolding.

Protect Your Skin

Portland cement is very alkaline, and prolonged contact with it can cause skin irritation and burns. Cement is also hydroscopic in that it tends to absorb moisture from your skin. Sand and fine aggregates can also be abrasive.

Clothing should not be allowed to become saturated with the moisture from the concrete because saturated clothing can transmit alkaline or hydroscopic effects to the skin. Waterproof gloves, a long-sleeve shirt, and full-length trousers should be worn.

If it is necessary to stand in fresh concrete while it is being placed, screeded, or floated, wear rubber boots high enough to prevent concrete from flowing into them.

When finishing concrete, waterproof pads should be used between fresh concrete surfaces and knees, elbows, hands, etc. Clothing areas that become saturated from contact with fresh concrete should be rinsed promptly with clean water to prevent continued contact with skin surfaces. Eyes or skin areas that come in contact with fresh concrete should be washed thoroughly with fresh water. Mild irritation of skin areas can be relieved by applying a lanolin cream to the irritated area after washing. Persistent or severe discomfort should be attended to by a physician.

When working with fresh concrete, begin each day by wearing clean clothing, and conclude the job or day with a bath or shower.

Other Safety Tips

Keep the work site organized and uncluttered. Injuries and spills happen when workers trip or stumble over tools and formwork lumber left scattered about the area.

If a power saw or other electric tool is being used, make certain it is properly grounded. Never leave power tools unattended.

It's almost impossible to do concrete and masonry work without drawing a crowd. Ask anyone not directly involved in the work to stay back a safe distance. Be particularly cautious with children. Never allow them to touch cement products or play around the work site.

Chapter 3
SOURCES OF CONCRETE

Concrete for your projects can be obtained in one of three ways:

- You can purchase fresh concrete from a ready-mix company that will deliver the mix by truck to your site.
- You can purchase aggregates and cement to mix your own concrete on site.
- Or you can simply use a packaged premix that only needs water.

The source you select will most likely depend on the amount of concrete you need.

The standard unit of measure in the concrete industry is the cubic yard. Although this might not sound like much, a cubic yard of concrete is quite a bit of material. Try this exercise.

Visualize a solid cube, 3' on a side. Now slice this cube into thirds along its length, width, and

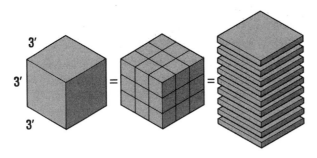

One yard of concrete is equal in volume to a cube measuring 3'×3'×3', or 27 smaller cubes 1'×1'×1', or 9 thinner slabs measuring 3'×3'×4".

height. You now have twenty-seven smaller cubes, each measuring 1' square. You could lay out these cubes to form a walkway 3' across and 9' long, but walkways are rarely 1' thick. In fact, in areas with good drainage and moderate climates, they can be as little as 4" thick. This means you could slice each 1' cube into three separate slabs, 4" thick. You now have enough material to increase the length of your walkway threefold, to 27'.

ESTIMATING CONCRETE NEEDS

When you estimate the amount of concrete needed for a given job, you're actually calculating the volume of concrete needed to fill the formwork. Most slabs, walls, footings, and the like, are nothing more than rectangular masses of concrete. To estimate the amount of concrete needed for these types of jobs, just multiply the thickness (height) of the project by its width and length, and then divide this number by 12. This determines the number of cubic feet of concrete needed. Divide the number of cubic feet by 27 to find the number of cubic yards needed. (See table 3–1.) For example, calculations for a concrete porch landing 6" thick, 12' wide, and 15' long are as follows:

TABLE 3-1: CUBIC YARDS OF CONCRETE IN SLABS AND WALLS					
Area in Square Feet	Thickness of Slab or Wall				
(L×W—slabs) (H×L—walls)	4"	5"	6"	8"	10"
25	0.31	0.39	0.46	0.62	0.77
50	0.62	0.77	0.93	1.24	1.55
75	0.93	1.16	1.39	1.86	2.32
100	1.25	1.55	1.86	2.48	3.10
200	2.50	3.10	3.72	4.96	6.20
300	3.75	4.65	5.58	7.44	9.30

Note: Does not include a loss factor due to uneven subgrade, spillage, etc.

$$\frac{\text{thickness (inches)} \times \text{width (feet)} \times \text{length (feet)}}{12} =$$

$$\frac{6" \times 12' \times 15'}{12} = 90 \text{ cubic feet}$$

$$\frac{90 \text{ cubic feet}}{27 \text{ cubic feet}} = 3.33 \text{ cubic yards}$$

Steps. When estimating concrete needs for steps, consider each step as an individual slab. In the example shown here, the bottom step will need 7-2/3 cubic feet and the top step requires 6 cubic feet for a total of 13-2/3 cubic feet without a loss factor.

Circular Forms. For circular or cylindrical forms, multiply the square of the circle's radius by 3.1416. This gives you the area of the form. Now multiply this area by the thickness or height of the form to determine its total volume.

Avoid Running Short

Running out of concrete in the middle of a project can be more than annoying—it can ruin your results. Always be sure you'll have enough concrete to completely fill the forms within a reasonable amount of time. The concrete in the forms must set up and age as a single mass, not as distinct layers or pockets. As will be explained later, some larger pours can be broken down

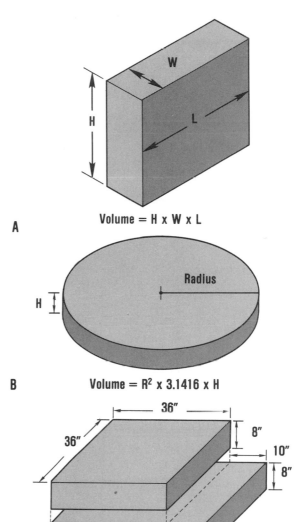

A   Volume = H x W x L

B   Volume = R² x 3.1416 x H

C

Calculating concrete volume for (A) square and rectangular forms, (B) circular forms, and (C) steps

into distinct sections, but again each section must be considered a single mass of concrete.

Once the pour begins, there will be no opportunity to run back to the building supply yard or phone in another order to the ready-mix company. So to be safe, measure and calculate your needs accurately, and then add 5 to 10% to the result to account for possible spillage, uneven subgrades, trial batches, and so forth. Batching and mixing your own concrete is further complicated by the fact that you have to estimate the correct amount of each ingredient needed: cement, coarse aggregate, and fine aggregate. Proportions of ingredients can also vary according to the strength and other properties desired. But before covering the details of proportioning and mixing your own concrete, let's take a closer look at the advantages and drawbacks of ready-mix and premix concretes.

READY-MIX CONCRETE

Ready-mix concrete is the most convenient and economical source of concrete for larger jobs. Having fresh concrete delivered to the job site that has been batched by professionals and thoroughly mixed during transit is a great advantage. It eliminates the work involved in buying, hauling, and mixing separate ingredients. Mistakes in proportioning ingredients also are eliminated.

Doing business with a local ready-mix company often makes their considerable expertise available to you at no extra cost. They know the local building codes pertaining to concrete mixes that can be used for certain types of work. Because of their experience with similar jobs in your area, ready-mix companies often can safely recommend a leaner, less expensive mix than you might otherwise feel comfortable in using.

In addition, ready-mix companies can accurately blend in admixtures to produce air-entrained concrete, fast- or slow-setting mixes, or colored concrete. You can rest assured the concrete will have the strength and workability characteristics you require.

Quantity Limits

Unfortunately, many ready-mix companies only deal in large quantities of concrete. The minimum amount many companies will deliver to a given site is 1 cubic yard. The reason for this is purely economic—one truck and driver can make only so many deliveries in a given day.

However, there might also be ready-mix companies in your area that target homeowners and smaller projects as their main customer base. These companies generally accept orders for less than 1 yard, and often offer metered delivery. With metered delivery, it is not necessary to estimate your exact needs ahead of time. A meter on the truck measures the exact amount of concrete deposited in your forms. That's the amount you pay for. Of course, it's always a good idea to calculate needs beforehand, particularly when comparing costs. Metered delivery also solves the problem of having too little or too much concrete, but such conveniences often carry a cost.

Some ready-mix companies also offer self-hauling service. With this system, you rent a small mixing trailer that is filled with quality mix at the plant, which is hooked to your car or truck using a universal trailer hitch. Trailers normally hold up to 1 cubic yard of concrete, which is mixed during the ride back to your job site. Some plain hopper trailers have no mixing capability and settlement of the coarse aggregate can occur during transport.

Be sure your vehicle can handle the weight of the trailer and the increased braking power

Ready-mix concrete is delivered to your site by a company specializing in batching concrete for specific types of projects and local conditions. It's the best source for large volume work.

needed. You won't have to worry about cracking sidewalks or other slab work with the weight of the trailer, but sinking in soft earth is still a concern. The trailer's dumping mechanism makes unloading into forms or wheelbarrows easy. You must clean out the drum or face an additional charge. Saturdays are prime time for rentals, so reserve the work date well in advance. Since you will be charged by the hour, or even half-hour, it's a good idea to assign one crew member the duties of picking up, washing out, and returning the trailer, while the other workers finish the concrete.

Check out several suppliers in your area before deciding which source of ready-mix is best for you. Compare such factors as minimum order amount, cost per cubic yard (or foot), cost for delivery, amount of time the driver may stay on site at no additional cost, cost for driver's additional time (if needed), and method of payment (cash, check, etc.). Don't be afraid to go with a company that appears to service mainly contractors if it offers the most advantages.

Access Problems

When the job site is near a good paved or dirt road, the driver should have little trouble maneuvering the truck into position for a chute-fed delivery. But troubles can occur when the truck must abandon the road to pass over sidewalks, driveways, and lawns. Overhead clearances to wires and between buildings can also pose problems.

Ready-mix trucks are tall, wide, and very heavy. A large truck can weigh in excess of 50,000 pounds, plus 4,000 pounds for each cubic yard of concrete on board. Such weight will certainly sink the truck into fresh, loose fill or muddy terrain. It can also damage older lawns, crack sidewalks and driveways, and break underlying sewer and drainpipes. Problems increase when the ground is saturated with rainwater. Laying down planking to distribute the load is futile when dealing with such weight. So, if ground conditions are soggy, it's best to postpone delivery until conditions improve.

Ask the ready-mix company for the dimensions of their trucks if you have doubts about adequate clearances. You'll only need several inches of overhead clearance, but it's best to keep the truck several feet away from buildings and structures. This is because even slight cross slopes or sinking to one side may tilt the tall truck significantly.

Finally, remember that the driver will judge where the truck can or cannot go safely. If you have doubts as to clearances or ground stability, have a representative from the ready-mix company visit the site before the delivery day.

Chutes, Conveyors, and Pumps. All ready-mix trucks are equipped with a delivery chute to help direct the flowing concrete into the forms. A typical chute measures about 10' in length, and some companies might provide extension chutes without charge. If you think extension chutes are needed, get a clear answer from the ready-mix company before placing the order. Some companies charge for extra chutes, and others do not offer them at all.

You can also build your own chute using a pair of 2×12 planks fastened side by side with cleats. Sides are 1×8 lumber. Be certain homemade chutes are strong and firmly supported. If there is a drop between the truck's main chute

Chutes are used to direct the concrete into the forms.

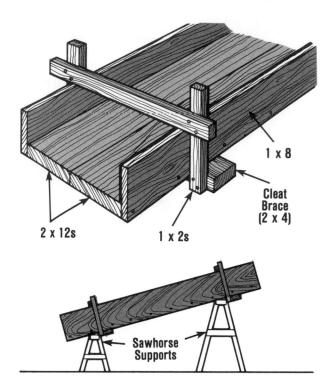

Details for building your own extension chute

and the second chute, the second chute will be subjected to strong downward force, so provide reliable support and brace the support and chute so neither can move.

Some companies offer conveyor delivery. A conveyor belt arrangement extends the natural reach of the truck and is extremely helpful in moving concrete uphill from the truck.

Another mechanical method of moving concrete is with a concrete pumper. The pumper consists of a truck with a hopper mounted on it that is connected to a gasoline-powered pump. As the ready-mix truck dumps concrete into the pump truck's hopper, the pump propels the mix through a wide hose running to the remote location. The hose end is easy to handle and can be directed into tight spots. Pump delivery is particularly good for deep forms such as higher walls. Pumping helps eliminate the shock of dropping concrete down a deep form. This shock can sometimes damage or displace forms.

Pumping range can be as great as 500' along horizontal runs. Pumping uphill reduces this range, so be sure to discuss the pumper's range

and capacity when ordering the service. Pumpers have a limit to the size of aggregate they can handle. Most can accommodate a maximum 1/2" stone, but others have a 3/8" limit. Make certain the specifications of your ready-mix concrete match the capacity of the pumper you are considering.

Concrete pumping services can be found in the phone directory. The ready-mix company you are ordering from might also be able to supply names of pumping services. Pump rates can be based on different formulas, such as yards pumped, yards and operator time, or yards, time, and mileage. All usually specify a minimum yardage, and all operators require facilities for hosing down their equipment after the job. Pumpers add another cost to the job, and coordinating the arrival of the pumper with the arrival of the ready-mix truck is another logistical headache. But for special cases, the pumper truck might be the only solution.

Wheelbarrow Hauling. In certain cases, there might be no alternative but to hand haul the concrete from the truck to the forms using a wheelbarrow. Use a contractor's wheelbarrow with an inflatable tire—small garden wheelbar-

Laying down a secure plank runway

rows aren't large enough or strong enough for this work.

Grease the axle beforehand and make certain the tire is inflated to maximum pressure. Unless you're wheeling over concrete, asphalt, or dry hardpan earth, you'll have to construct a runway of planking to avoid sinking into the soft ground. Tie the planks together with wood strips as shown and be sure all joints are level and properly supported. Even a slight bump can throw a wheelbarrow loaded with 150 pounds of concrete out of balance.

Figure on hauling 1 cubic foot of concrete per trip. That's close to thirty trips per yard of concrete, so have several strong crew members ready to do some hard work. A two wheelbarrow setup is highly desirable for larger pours, but always exhaust all other methods of moving concrete before resorting to hand hauling.

Ordering Ready-Mix

Contact the ready-mix supplier at least one week in advance if you want to be reasonably sure of delivery at the time of your choice. As mentioned in chapter 2, early morning delivery provides maximum working time and optimum temperatures during summer months, but everyone will be asking for these choice times so call early and be flexible. Calling early will also give the supplier time to send someone out to check your work site if there is a question concerning access or clearances.

Since there's no way to accurately predict the weather a week or more in advance, ask the company what their policy is concerning cancellation due to bad weather. Most require advance notification, so you'll have to watch the weather reports as the delivery day nears. You obviously don't want to work in a steady rain or freezing temperatures. Rain during the critical finishing process can ruin the concrete, but a shower several hours after the concrete has gained its final set can actually help the curing process.

When ordering, specify the amount needed in cubic yards and fractions of yards, such as 2 yards, 4 cubic feet, or 1-1/2 yards, etc. State the maximum coarse aggregate size required, based on the guidelines provided on pages 34 and 35.

Specify the type of project, such as a sidewalk, steps, wall, basement floor, driveway, porch slab, foundation, and so on. Once the ready-mix company knows the application details, they can safely recommend a suitable compressive strength and cement content for local conditions. Don't hesitate to talk about strength specifications, and always specify air-entrained concrete for exterior work.

Confirm the need for extension chutes or conveyors and their cost. You should also know the exact amount of time the driver is allotted at the site before overcharges begin.

Preparing for Delivery

Building forms on the day of delivery is risky business. Have all formwork ready a day or so in advance and double check to make certain the forms are sturdy and placed exactly where you want them. Make certain the subgrade is properly prepared and any crushed stone or iron rebar or mesh is properly positioned.

Get your crew on the job site early and run through the delivery process ahead of time. Plan where the truck will back up and which forms will be filled first. Assign specific tasks to each crew member. If you have to spread the concrete over a large area, be sure to have a shovel

On large jobs, such as this patio, crew members should be equipped with a shovel for spreading concrete. Protective boots are also a must.

or strong rake for each crew member. Set up planking for wheelbarrow hauling if needed. Double check that all screeding, floating, and finishing tools are on hand.

Accepting Delivery

The sight of a massive ready-mix truck pulling up to your job site can be a bit intimidating, but if you've done your planning and preparation things should go smoothly. Take charge of the operation. Note the truck's time of arrival and direct the driver to the preselected unloading point.

The truck is equipped with a water tank so the driver can add water to the mix. Most mixes are batched a little stiff at the plant with the driver adding water to reach the final consistency needed for the job. Moving concrete over a large area such as a slab might require a slightly wetter mix, but add water sparingly in small increments. Remember that water cannot be removed once added, and there are no dry materials on the truck to firm up the mix if a mistake is made.

The driver will stand at the rear of the truck and release the concrete as per your directions. Make certain the driver can see and understand your hand signals. Remember to account for the concrete in the delivery chute when timing the closing of the mixing drum door.

Don't count on the driver picking up a shovel and lending a hand. That's not part of his or her job, although some might do it if they see you're in trouble. Also, don't let the driver dump and run. Discharge the concrete at a rate the crew can handle at a steady, not frantic, pace. Keep an eye on your watch and know where you stand in relation to your allotted delivery time. A driver running behind schedule might try to cut you short, but remember the price of the concrete includes a certain set delivery period. Insist on it if needed.

Make certain all forms are fully filled and even a little overfilled. If the forms are deep, such as those for a retaining wall, poke a board or pole into the mix every 2' or so to make sure no air

pockets have formed and that the concrete has spread into every corner of the mold.

Extra Ready-Mix

The task of accurately estimating concrete needs is particularly important when purchasing ready-mix. Metered delivery, as mentioned earlier in this chapter, might not be available, or its cost might be prohibitive for bigger jobs.

Obviously, you want enough concrete delivered to do the job without paying for extra materials you can't use or dispose of. For example, 10% extra on a 2-yard order is more than 5 cubic feet of concrete. With accurate measurements and calculations you might get by with a 3 to 5% extra cushion. You should also always have some type of filler on hand to add volume to the ready-mix if needed. Adding a bucket or two of crushed stone, brick, or old clean concrete rubble to the ready-mix will make up for a small shortfall without adversely affecting the new concrete. But don't overdo it, or you'll change the character of the mix.

If there is a substantial amount of concrete left in the truck after the forms are full, you have two options. The driver can dump it in a separate area so it can be used for another small project, or he or she can take it back to the ready-mix

The excess concrete from the large patio shown earlier was used to fill these paving block forms. Plastic sheeting was used to keep the base smooth and clean.

plant so you won't be faced with the problem of disposal. But remember, either way you still pay for the amount of concrete listed on the original order.

Examples of small side projects suitable for extra ready-mix are setting a post or two, placing a small slab under an outdoor faucet, or casting a molded planter, downspout splash guard, or wheel chock. But don't allow a side project to syphon materials and attention away from the main job. Accept the fact that there might not be any extra concrete and your side project preparations will have been done for nothing.

PACKAGED PREMIXES

Although ready-mix is the ideal answer when large amounts of concrete are needed, packaged premixes are the most convenient solution for small jobs or situations where buying, handling, and storing separate ingredients pose problems.

Premixes contain the proper blend of cement and aggregates in the same bag. You simply add water as directed to produce a mix with a plastic-like consistency. The coarse aggregate in premixed concrete is normally only 1/2" in diameter, but this small size makes mixing, placing, and finishing easier.

Premixed concrete is available at home center stores, building supply yards, hardware stores, and even some supermarkets. Packages are sized by weight; 45, 60, 80, and 90 pounds are common sizes. The contents of a typical 60-pound bag will mix up to produce about 1/2 cubic foot of concrete.

At 1/2 cubic foot per bag, you can easily see that the number of bags needed for larger projects quickly adds up. Premixes also cost more than buying and mixing your own ingredients. But premixes also offer distinct advantages. For example, you'll be handling one bagged product instead of several loose aggregates plus cement. All proportioning except the water content has been done for you. You're assured of quality ingredients when you buy reputable brand names. And as long as you keep it dry, any unused premix can be stored for future projects. There will be no small piles of unused sand and aggregate scattered about your property at the end of the project.

Follow the premix manufacturer's directions regarding mixing and use. Always set some premix aside so you'll have material to add to the mix if you make the first batch too soupy. Premixes are batched and packed at regional plants, and the cement content (strength) and aggregate type are selected to meet regional conditions and availability. However, premixed products are rarely air-entrained. This is because hand mixing is not rigorous enough to generate the tiny air bubbles within the concrete.

With a power mixer, you can probably batch up premix faster than mixing separate ingredients, so the size of the projects you tackle with premix will largely depend on how heavily you weigh convenience against cost. Premixes have traditionally been used when several cubic feet

Packaged premixes contain the proper blend of cement and aggregates. Simply add water and mix.

of concrete are needed, such as when setting posts, casting molded projects, or constructing small service walls, entrance landings, mounting slabs, small footings, short curbs, or stoop steps.

Larger premix jobs are certainly possible, particularly if you break down the work into smaller sections. A patio with a permanent grid of treated lumber is a good example. Grid sections can be filled one at a time, and you can do as many as you can handle in a given work session. To make certain each section cures to the same shade, keep the amount of water used per bag constant throughout the project and use the same curing technique on all sections. If you have the room to safely store the bags, purchase all you need in one visit and ask for packages from the same shipment. This increases the likelihood that all packages are from the same regional plant batch, minimizing slight differences in ingredients and proportioning that could lead to color variations.

Types of Premixes

Premix manufacturers offer a number of basic products, plus a variety of specialty and repair products. All use cement as their main bonding agent, but aggregates and additives are varied to meet specific needs.

Concrete Mix. Concrete mix, also called *gravel mix*, contains cement, sand, and gravel. It is used for all general concrete work.

Sand Mix (Fine Topping). This mix of cement and fine sand is designed for jobs where concrete is needed in less than a 2″ thickness. Sand mix cures to a stronger finish than ordinary premix concretes because of its higher cement content. It can also be used for filling cracks, resurfacing, stuccoing, and laying flagstone and paving brick.

Fast-Setting Premix. This fast-setting mix is ideal for setting posts and poles and for slab work 2″ or thicker. Initial set can be in as little as fifteen minutes with walk-on time in less than one hour.

Surface-Bonding Mixes. These are specialty mixes containing additional bonding agents or fiber-reinforced materials. Applications vary among specific products, but typical jobs include top bonding to damaged walls and slabs, bonding together dry stacked masonry block, or paraging over rigid insulation board.

Prepackaged mortar mixes, stucco finish coatings, waterproofing coatings, anchoring cements, water-stopping cements, and specialized repair products are all readily available. Their applications and use are covered in more detail in later chapters.

MIXING CONCRETE FROM SCRATCH

The third source of quality concrete is to take matters into your own hands and mix it yourself. Mixing from scratch requires a little more calculating and measuring, but the problems are not insurmountable.

Scratch mixing is common for small to medium-size projects requiring between 1/2 and 2 yards of concrete. Working at a steady pace, a two-person team using a medium-size power mixer should be able to mix, place, and finish between 1 and 2 yards of concrete per day.

Scratch mixing is less expensive than using premixes. With premixes, the maximum aggregate size is usually 1/2″, but scratch mixing lets you select larger coarse aggregate sizes. You can also alter the mix proportions to fit the job at hand with the most economical mix possible.

Tips for buying and selecting aggregates are given in chapter 1. When ordering, always add at least 10% to your estimates to allow for waste and error. Much of the labor involved in scratch mixing results from having to move the aggregate and cement about, so carefully plan the drop point. It should be as close to the mixing setup as possible, which in turn should be as close as possible to the place where the concrete will be used. If your property is pitched, it is much better to unload above the job site rather than below it. It's a good idea to build a short, temporary retaining wall of planks or concrete block and dump the sand and aggregate against it. The wall will keep the materials from spreading out in an ever-increasing circle as you dig into them with your shovel. If the building supplier is delivering the aggregates, make it a point of being there during unloading. The driver might not take the time to analyze the best drop site, and once the dump is made you'll have to work around his or her judgment.

Mixing Proportions

As discussed in chapter 1, the strength of concrete depends mostly on the proportions of the ingredients. The most accurate way to mea-

A workable mix contains correct amounts of cement paste, sand, and coarse aggregate. With light troweling, all spaces between coarse aggregates are filled with sand and cement paste.

A wet, soupy mix contains too little sand and coarse aggregate for the amount of cement paste present. Such mixes are costly, with low durability and a tendency to crack.

A stiff, harsh mix contains too much sand and coarse aggregate for the amount of cement paste present. Harsh mixes are difficult to place and finish smooth.

Another example of a stiff, harsh mix containing too much coarse aggregate and not enough sand. This mix would result in honeycombed, porous concrete.

A sandy mix lacks sufficient coarse aggregate. It is easy to place and finish, but quite expensive. Such a mix can be used as a topping coat if properly proportioned, but applications thicker than 2" will likely crack.

sure and proportion the aggregates, cement, and water in a mix is by weight. Proportioning ingredients by volume is much more common, however, and is accurate enough for jobs around the home.

TABLE 3-2: PROPORTIONS BY WEIGHT TO MAKE 1 CUBIC FOOT OF CONCRETE FOR SMALL JOBS

Maximum-Size Coarse Aggregate (in.)	Air-Entrained Concrete				Nonair-Entrained Concrete			
	Cement (lb)	Wet, Fine Aggregate (lb)	Wet, Coarse Aggregate (lb*)	Water (lb)	Cement (lb)	Wet, Fine Aggregate (lb)	Wet, Coarse Aggregate (lb*)	Water (lb)
3/8	29	53	46	10	29	59	46	11
1/2	27	46	55	10	27	53	55	11
3/4	25	42	65	10	25	47	65	10
1	24	39	70	9	24	45	70	10
1-1/2	23	38	75	9	23	43	75	9

*If crushed stone is used, decrease coarse aggregate by 3 lb and increase fine aggregate by 3 lb.

Proportioning by Weight. Table 3–2 lists suggested proportions by weight to make 1 cubic foot of concrete. You'll notice that if you divide the weight of the water by the weight of the cement used, a rather low water–cement ratio of about 0.35 is used. This is because the actual water–cement ratio is somewhat higher when you account for the moisture commonly present in wet sand and aggregates.

To estimate individual material needs, multiply the values listed in the table by the number of cubic feet of concrete needed. For example, if 2/3 of a cubic yard or 18 cubic feet are needed for a slab using 1" maximum aggregate, calculations based on table 3–1 are as follows:

Cement: 18 × 24 lb =
432 lb ÷ 94-lb bag = 4.6 or 5 total bags
Sand: 18 × 39 lb = 702 lb + 10% =
772 lb total

Coarse aggregate: 18 × 70 lb =
1,260 lb + 10% = 1,386 lb total

Water: 18 × 9 lb = 162 lb ÷ 8.33 lb/gal =
19.45 gal

Aggregates are sold by the ton (2,000 pounds) or by the cubic yard. To convert from pounds to cubic yards or vice versa, use an estimated weight of 90 pounds per cubic foot of sand (wet) and 100 pounds per cubic foot of coarse aggregates.

So the 775 pounds of sand needed converts to 8.6 cubic feet or about 1/3 of a cubic yard. Coarse aggregate volume is approximately 14 cubic feet or a little over 1/2 cubic yard.

A

B

C

(A) Wet sand, which describes most sand, forms a ball when squeezed, but leaves no noticeable moisture in your palm. (B) Damp sand crumbles when squeezed into a ball. (C) Very wet sand forms a tight ball and leaves moisture on your palm. Standard proportioning is based on wet sand; water added to the mix may increase or decrease when damp or very wet sands are used.

Maximum-Size Coarse Aggregate (in.)	Air-Entrained Concrete				Nonair-Entrained Concrete			
	Cement	Wet, Fine Aggregate	Wet, Coarse Aggregate	Water	Cement	Wet, Fine Aggregate	Wet, Coarse Aggregate	Water
3/8	1	2-1/4	1-1/2	1/2	1	2-1/2	1-1/2	1/2
1/2	1	2-1/4	2	1/2	1	2-1/2	2	1/2
3/4	1	2-1/4	2-1/2	1/2	1	2-1/2	2-1/2	1/2
1	1	2-1/4	2-3/4	1/2	1	2-1/2	2-3/4	1/2
1-1/2	1	2-1/4	3	1/2	1	2-1/2	3	1/2

TABLE 3–3:   PROPORTIONS BY VOLUME* OF CONCRETE FOR SMALL JOBS

*The combined volume is approximately 2/3 of the sum of the original bulk volumes.

Bathroom scales are accurate enough to weigh materials. Each ingredient should be weighed in its own container. Five-gallon metal or hard plastic buckets are ideal, but make sure the handles can hold the weight. Remember to account for the weight of the container when measuring out the materials.

Weigh out a practical amount of each ingredient, then mark the level on the inside of its container. Future batches can be measured by filling to this mark. The scale is no longer needed unless the aggregates change or their moisture content is altered.

Proportioning by Volume. Proportioning by the bulk volume of materials is the most popular method of batching concrete. Experienced concrete workers often do a good job measuring by the shovelful, but using containers is still recommended. Table 3–3 lists proportions by volume for various aggregate sizes. Using the same example as above, you can see that table 3–3 lists the proportions for air-entrained concrete using 1" maximum aggregate as 1 part cement, 2-1/4 parts sand, 2-3/4 parts coarse aggregate, and 1/2 part water. Once again the water content is based on the use of wet sand.

A "part" can be any volume you select as long as you remain consistent. Use a container size you can lift and load without struggling. It's useful to remember that each 94-pound bag of cement contains 1 cubic foot of material, and that 1 cubic foot equals about 7-1/2 gallons.

Concrete Volume. The final volume of the concrete mixed will not equal the combined volume of the individual ingredients. It will normally compact to about 2/3 of the original bulk volumes. Again consider the 1" aggregate mix. If you start out using a 1 cubic foot bag of cement, the bulk volumes are as follows:

Cement: 1 cubic foot

Sand: 2-1/4 cubic feet

Coarse aggregate: 2-3/4 cubic feet

Water: 1/2 cubic foot

Total bulk volume:
6-1/2 cubic feet × 2/3 =
4-1/3 cubic feet final concrete volume

When machine mixing, proportioning volumes will be limited by the maximum batch capacity of the machine. Obviously you'll want to mix the largest batch possible without overloading the mixer. Don't confuse batch capacity with drum volume. Mixing capacity is usually only about 60% of the total drum volume. So a 7 cubic foot drum can mix about 4 cubic feet of concrete at a time, and so on. Maximum batch size is usually stated on the mixer's name plate. If you're renting a mixer, know its capacity before leaving the rental yard.

Water Content. Adding the right amount of water is the most difficult aspect of mixing from scratch. Dry sand is rarely available for concrete work. Sand used on most jobs contains some moisture, which must be accounted for as part of the mixing water. Check the water content of sand by squeezing the sand with your hand. Wet sand forms a ball and leaves no noticeable moisture on your palm. Damp sand falls apart when squeezed and released. And very wet sand forms a ball when squeezed and leaves moisture on your palm.

As mentioned earlier the water values listed in the proportioning tables in this chapter are based on the use of wet sand. If you are measuring by weight using table 3–2 data, you can make an accurate adjustment for damp or very wet sand. If the sand is damp, decrease the quantity of sand listed by 1 pound and increase the water quantity listed by 1 pound. If the sand is very wet, increase the sand and decrease the water content by the same 1 pound each.

When proportioning by volume, adjustments for damp or very wet sand are less precise but should still be made. Trial batches are a must when mixing your own concrete and guidelines for adjusting mix problems are listed in the next section. The final word on water content is always the water–cement ratio. A gallon of water weighs about 8-1/3 pounds, and a cubic foot of water weighs almost 62 pounds. Combining one 94-pound bag of cement with a total of 6 gallons of water (50 pounds) results in a water–cement ratio of 0.53. This is about the highest ratio recommended for home use. In fact the closer you mix to 5 total gallons of water per bag (water–cement ratio 0.45), the stronger your concrete will be.

Alternate Proportions. Tables 3–2 and 3–3 list proportions for the strongest practical mixes using various aggregate sizes. These proportions should be used for essentially watertight concrete exposed to outside weather, deicing agents, and the like. Use these proportions for all concrete used in walks, driveways, patios, swimming pool decks, floors, and so on.

By increasing aggregate content, less expensive but slightly weaker mixes are possible. For example, a 1:2-3/4:4 mix produces moderately strong concrete useful for most foundations, walls, and structures not directly exposed to the weather. A 1:3:5 mix can also be used for heavy, massive foundations, thick retaining walls, and masonry backing. Once again, the slab or wall thickness controls the maximum coarse aggregate size used on the job. And none of these

TABLE 3-4: MATERIALS TO MAKE 1 CUBIC YARD OF CONCRETE

1:2-1/4:3 Concrete

Materials	Quantity
Cement	6 bags
Sand (wet)	14 cubic feet or 0.51 cubic yard
Coarse aggregate	18 cubic feet or 0.66 cubic yard
Water	36 gallons

1:2-3/4:4 Concrete

Materials	Quantity
Cement	5 bags
Sand (wet)	14 cubic feet or 0.51 cubic yard
Coarse aggregate	20 cubic feet or 0.74 cubic yard
Water	30 gallons

1:3:5 Concrete

Materials	Quantity
Cement	4.5 bags
Sand (wet)	13 cubic feet or 0.48 cubic yard
Coarse aggregate	22 cubic feet or 0.8 cubic yard
Water	27 gallons

alternate mixes should be used with aggregate having a maximum size greater than 2-1/2". Water content should never exceed 6 gallons per 94 pounds of cement used. Table 3–4 lists the approximate materials needed to mix 1 cubic yard of concrete using the three most popular mix proportions.

Machine Mixing Trial Batches

The best way to mix concrete is with a gasoline or electric power mixer. It guarantees thorough mixing and is the only way to produce air-entrained concrete on site.

With the mixer stopped, begin by loading all of the coarse aggregate and half of the mixing water. If a liquid air-entraining agent is being used, add it as a portion of this mixing water. Now start the mixer and add the sand, cement, and remaining water. Once all ingredients are loaded continue mixing for at least three minutes. When properly blended, the fresh concrete will have a uniform color. Streaking indicates dry patches in the mix.

Place the concrete in the forms as soon as possible after mixing. If the concrete shows signs of stiffening in the drum, remix it for about two minutes to restore its workability. If remixing does not result in a plastic mix, discard the concrete. Never add water to concrete that has stiffened to the point where remixing does not restore workability.

Mix a trial batch of concrete using the proportions provided in the tables. Remove a small sample from the mixer and test it for stiffness and workability. A good mix will be wet enough to stick together without crumbling. It should be plastic enough to slide down the surface of a shovel or trowel, not drip off like a liquid.

Next flatten the mound slightly to form a plateau. When worked with a trowel, there should be sufficient cement paste to fill in all voids and create a smooth finished surface.

A wet mix that cannot hold its shape has too much cement paste and too little aggregate. To salvage the trial batch add more sand and coarse gravel or stone to the mixer and remix for a minute or so. Repeat until the proper consistency is reached. For future batches, return to the original amounts of fine and coarse aggregates, but decrease the amount of mixing water used.

Stiff mixes have the opposite problem, insufficient paste to fill in spaces between the aggregates. In future batches decrease the amount of fine and coarse aggregate, but use the original amounts of cement and water listed. To save the stiff trial batch, add a cement paste mixed to a proportion of two parts cement to one part water by weight. Never add water alone to a mix that is too stiff.

If the mix is too sandy, decrease the amount of sand and substitute an equal amount of coarse aggregate in future batches. With a stony mix, cut back on coarse aggregate and increase sand content.

Record all adjustments to ingredient amounts on the batching cans. Once you've hit upon the correct mix, you should not have to adjust proportions for the remainder of the work session.

The mixer must be thoroughly cleaned immediately after the last batch of the day. Scour the mixer by running it with several shovelfuls of coarse aggregate and water. A vinegar solution will help remove light cement films from the exterior of the drum. To remove concrete buildup on the inside of the drum and blades, scrape with a flat piece of lumber, and then scrub with a wire brush. Don't try to chisel or hammer the concrete free; it's easier than you might think to damage the drum or blades. Remove stubborn buildup with a solution of one part hydrochloric acid (muriatic acid) in three parts of water. Allow thirty minutes for penetration, then scrape or brush the buildup and rinse with clear water.

WARNING

Hydrochloric acid is hazardous and toxic. Avoid skin contact and fumes. Wear rubber or plastic gloves and chemical safety goggles. If the acid is used indoors, provide adequate ventilation.

Dry the mixer drum thoroughly to prevent rusting, and store the mixer with the opening of the drum pointing down.

Hand Mixing

For small jobs requiring a volume of concrete less than a few cubic feet, it is sometimes more convenient and less expensive to mix by hand.

Hand mixing is not vigorous enough to make air-entrained concrete, even if you use air-entraining cement or an air-entraining agent. Hand mixing, therefore, should not be used for concrete exposed to freezing-thawing conditions or deicers.

When performed at a comfortable pace and using the correct technique and equipment, hand mixing is not exhausting work. Some tips for hand mixing are:

- Gloves that slide on the hoe handle waste more energy than is worth the protection they provide. Unless your gloves give you a firm grip, mix with your bare hands.
- Prevent blisters from forming on bare hands by keeping your hands dry at all times and shifting hand positions continu-

ously so the tool handle does not press and slide across the same fold of skin.

- If your hands become sore, stop working, get relief, or protect the sore areas with several layers of athletic tape before blisters have a chance to form.
- Wear heavy-soled shoes and thick, protective socks. If your feet slip and slide as you push back and forth, this wastes energy and leads to blisters.

Mixing in a Mortar Box. A mortar box is an excellent container in which to mix concrete. Place the box on a level surface and wet down its inside surface. Spread out all the coarse aggregate in an even layer in the bottom of the box. Leave about 1/4 of the box clear to act as a mixing well. Spread the sand component of the mix on top of the coarse aggregate and follow with the cement portion of the mix. You now have a three-layer sandwich of stone, sand, and cement in the proportions called for in your mix formula.

Pour a portion of the mix water into the well area of the box. Using a hoe, slice down through the sandwich and pull a small cross section of dry materials into the water. Turn it over several times to blend thoroughly and then cut off another small section. Continue working in this way until the mix in the well area of the box reaches the proper consistency for placement. You can then remove and place this concrete or add more water and continue to work forward through the entire sandwich of dry materials. You should only be mixing a very small amount

of the overall materials at a given time. When you finish, the mixing well will be at the opposite end of the box from where you started.

Mixing in a Wheelbarrow. You can't use the mortar box technique in a wheelbarrow. There just isn't enough space. Instead, spread the sand on the bottom of the wheelbarrow following with layers of coarse aggregate and cement. Add water in increments and mix with a hoe or shovel. Hold the hoe short to help control it, and don't lift a shovel any higher than needed to clear it when you turn it sideways. Limit amounts to 1 to 2 cubic feet at a time or mixing will be cramped and difficult.

Mixing on Flat Surfaces. This technique is similar to the mortar box principle except that the materials are positioned in a ring on a hard, smooth surface such as a concrete floor or driveway or a heavy sheet of plywood. The center of the ring acts as the mixing well. The ring is layered with coarse aggregate on the bottom followed by sand and cement layers. Water is added to the center of the ring and ingredients are pulled into the mixing areas using a hoe. As the mix reaches the proper consistency, it is removed for immediate placement.

For best results, wet down the surface before forming the rings. This helps with cleanup. Also, keep overall ring height below 6" or materials will become hard to handle.

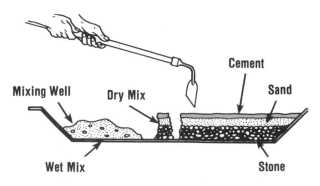

Layered mixing in a mortar box

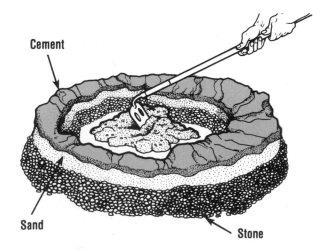

Layered ring technique used for mixing on a flat surface

Chapter 4

SITE PREPARATION AND FORMWORK BASICS

A well-prepared subgrade and good form-work stage a successful concrete project. Preparation can be broken down into three distinct steps:

1. Correctly locating the position of all formwork, trenching, and so forth
2. Digging and excavation to produce a level, even subgrade
3. Construction and bracing of the necessary forms

Laying out a small slab project on level ground using a framing square and tape measure

LAYOUT AND LOCATION

The dimensions and shape of your project must be carefully marked out on the work site before the first spade of earth is moved or the first nail is driven. Footings, foundations, and walkways must run straight and true. Most slabs are designed to be square or rectangular in shape. Curved formwork and circles must be smooth and symmetrical.

Small slabs and walkways located on level sites can be accurately laid out using a large framing square and tape measure. But for any project measuring more than a few feet on a side and for all projects on hilly or rolling terrain, it's essential to check dimensions using some basic geometry.

The 3-4-5 Principle

The most precise method of squaring corners is to use the 3-4-5 principle of right triangles. As shown in the illustration, if one leg of a right triangle measures 3' and the other leg 4', the hypotenuse, or longest leg, will measure 5'. This principle holds true for multiples of this ratio such as 6-8-10 and 9-12-15.

To lay out a large square or rectangular area you'll need some sturdy wooden stakes, a roll of heavy string or twine, a line level, a measuring tape, some tacks, and a hammer. Layout work is definitely a two-person operation, so you'll need a helper.

The 3-4-5 principle for establishing square corners

On small jobs, locating a stake at each corner of the site and running lines between them will likely be accurate enough to provide a guide for removing sod and topsoil, and positioning forms. Measurements and squareness can be easily rechecked during final formwork construction.

For large layouts, it's best to position the stakes slightly back for the actual corners of the work site. When this is done, the true corners of the work site are located at the points where the string lines intersect. Setting the stakes back from the corners allows for the most precise measurements possible. It also gives you room to construct the forms and prepare the subgrade without disturbing the layout lines.

Remember that all lines strung between the stakes must be level with one another for accurate measurements. If the site has a definite slope to it, begin at the highest corner or make sure your stakes are long enough to account for the rise in slope. Drive stake No. 1 into the ground about 1' back for the desired final location of corner No. 1. Measure out past corner location No. 2 and drive stake No. 2 into the ground. At this point in the layout, don't drive the stakes in too solidly. You might have to move them to make adjustments.

Run line A from stake No. 1 to stake No. 2. Hook the line level to line A to make sure it is running level. Then precisely locate the final locations or corners No. 1 and No. 2 by measuring carefully along line A and pushing tacks through the line at these points.

You can now locate stakes No. 3 and No. 4 and run line B between them. Line B runs perpendicular to line A, passing directly over the tack marking the location of corner No. 1. To ensure that the angle between lines A and B is a true right (90°) angle, the 3-4-5 principle is used to check it.

Measure out from corner No. 1 along line A a distance of 3' and mark this point in the line by passing a tack through the string. Next, measure out from corner No. 1 along line B a distance of 4' and mark this point in the line with a tack.

Now measure the distance between the tacks. If line B is perpendicular to line A, the distance between the tacks will be 5' exactly. If it isn't, adjust the position of line B or line A until the 5' distance is obtained. You can now drive all stakes firmly into the ground to secure corner position No. 1.

Use the same technique to lay out lines C and D and establish final positions for corners No. 3 and No. 4. Remember to keep all lines

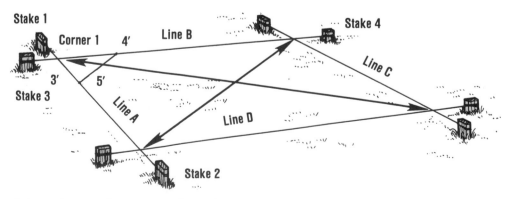

Set stakes back from the actual corners on larger layouts.

running level to one another or the measurements will not be accurate.

As a final check, measure the distance between opposing corners. These diagonal measurements will be equal if the layout is square at all corners. If the diagonals are not equal, recheck all measurements and run the 3-4-5 test at each corner. For bigger jobs, a 6-8-10 or 9-12-15 ratio will give more accurate results.

Batter Boards

Batter boards are normally used when digging deeper trenches in preparation for a foundation or wall, but they can also be helpful for slab work.

Batter boards are set back from the actual work site so they will not be disturbed by construction work. They serve as a record of the important job measurements, such as trench and footer dimensions, the location of the outer wall edge, and so on. Construct batter boards using the following procedure:

1. Lay out the perimeter of the project as described above. Drive in stakes at each corner.
2. About 5′ outside of each corner, drive three 1×4 stakes into the ground as shown and nail the batter boards to them. Since a typical footer trench must

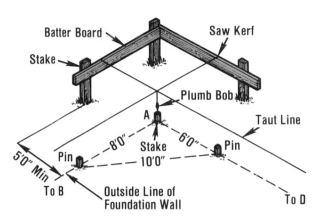

Position string lines on the batter boards using a plumb bob suspended over the true corner position.

be at lest 3′ wide to allow room to work, the batter boards should be 4′ to 5′ long.

3. Transfer the dimension lines of the project to the batter boards by hanging a plumb bob over the outer edge of each corner stake and stretching a line between the batter boards. This is one job where a helper is essential.

4. Now that the outside corner dimensions of the project are accurately marked on the batter boards, you can use them as a reference point for locating other important dimensions such as the position of the trench, formwork location, and footer edges. Mark these positions on the batter boards so that you will be able

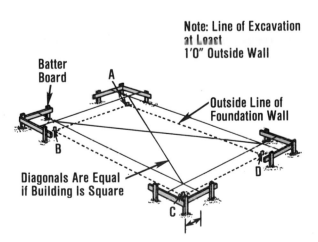

Batter boards are constructed a short distance back from the corners of the slab or footer trench.

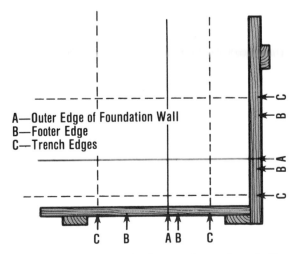

Typical dimensions marked out on batter boards. Use a small saw kerf to hold the line in position.

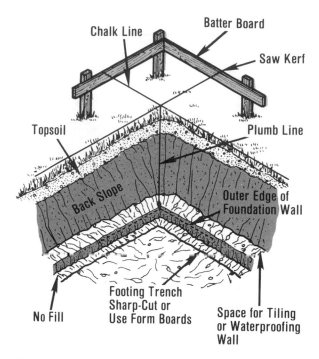

Transferring dimensions from the batter boards to the trench or footer using a plumb bob

Site preparation begins by removing all sod, roots, and organic material down to a firm earth base.

to quickly and accurately string lines between the batter boards. Cutting a thin saw kerf at these positions is a good method of holding lines in position.

5. To transfer construction dimensions from the batter boards to the work, string lines between the appropriate marks on the batter boards and use a plumb bob to transfer locations to the base of the trench, footer, and so on.

PREPARING THE SUBGRADE

Preparation of the subgrade, or the ground on which the concrete will rest, is critical to your project's success. This is particularly true of slab-type projects such as walks, driveways, patios, and foundations. A poorly prepared base can result in slab settlement, cracking, and even structural failure.

For best results, the subgrade should be fairly hard, uniform, and level. It should also provide good water drainage away from the concrete.

Excavating the Site

Once the site has been laid out, excavation can begin. Start by removing all grass, sod, and roots from the construction area. The depth of the excavation can vary a great deal, depending on the slope of the land, the thickness of the slab being built, the local climate, and the soil and drainage conditions encountered. Excavating deeper trenches for footers and foundations is covered in chapter 7. This chapter concentrates on site preparation and formwork for slab-type work.

Keep in mind that undisturbed soil supports a concrete slab better than soil that has been dug out and replaced with poorly compacted fill. So only do the digging needed to bring the site to the required level and grade. In temperate climates free of ground heave and erosion, removing the sod and compacting the top soil might be all that is required. However, it's a good idea to remove at least 2″ of soil in most cases. This will set the top edge of a 4″ slap about 2″ above ground level.

The subgrade must be uniform throughout to correspond with temperature and moisture changes. Level the earth surface so there are no deep holes or protruding objects. Fill in any soft spots or holes with sand, gravel, or crushed stone. Thoroughly compact the soil with a tamper to make it firm and uniform.

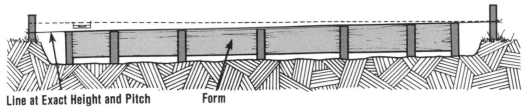

Line at Exact Height and Pitch   **Form**

Setting formwork pitch over a long distance using a line level

Pitch Considerations. A concrete slab must be pitched, or angled at a slight incline, to provide for necessary drainage. Pitch must angle away from a house or other structure, such as a garage. For most walkways and smaller slabs the needed pitch can be built into the formwork by setting one side of the form slightly higher than the other. But for larger projects, such as a patio or driveway, the pitch angle must be designed into the subgrade.

Proper pitch is usually 1/4″ per foot, but slabwork in areas subject to heavy rainfalls might require greater pitch. Pitch specifications for slab work is usually spelled out in local building codes.

Based on a 1/4″ per foot specification, a patio slab that extends 20′ out from a home should have a total pitch of 5″. That is, the far edge of the patio should be 5″ lower than the side that butts against the house.

As shown in the illustration, you can set and check pitch using a line level. Begin by marking the proposed height of the slab on the high-side layout stake. The top of the formwork will be set at this level. Run a level line to the low-side layout stake and transfer this high level mark to the low-side stake. Next, measure down from this mark the required pitch distance, and lower the line to this point. The formwork at the low-side stake will be set at this level.

Remember, the pitch will affect the angle of the slab, not its thickness. Be sure your excavation reflects this gradual pitch, or the finished slab might have thin and thick areas that will affect its stability and strength.

Drainage. In areas of poor drainage or unstable soil, a layer of crushed stone or gravel should be placed on top of the prepared base. This gives the slab a stable base and allows

Adding a gravel or crushed stone base for better drainage. The screed board in front of the shovel is used as a quick visual check of the stone layer thickness.

groundwater to drain from under the slab more easily. The thickness of the crushed stone or gravel layer is usually 4″ to 6″. Local building codes might specify guidelines for larger projects.

The crushed stone or gravel is normally set in place after the formwork has been completed, but you must account for the depth of this layer when excavating the site and establishing pitch and formwork positions. For example, if you plan to use a 4″ layer of crushed stone under a 4″-thick walkway, you'll have to excavate to a depth of 6″ if you want the top of the walk to be 2″ above ground level.

On larger projects, the layer of gravel or stone should follow the pitch of the subgrade. And to keep water from collecting under the slab, the bottom of the subgrade should not be lower than the surrounding finished grade.

A drag board notched to fit over the formwork provides an easy way to level the stone or

Using a notched screed board to level the layer of gravel or stone to the desired thickness

gravel to the proper height. The gravel or stone should be tightly compacted and, if possible, extend beyond the edges of the formwork for several inches in all directions. This overhang will help keep water from collecting against the slab edge.

FACTS ABOUT FORMS

Forms for concrete range from simple to elaborate, but forms for any concrete project must be tight, rigid, and strong. If forms are not tight, there will be a loss of concrete, which might result in honeycomb or a loss of water that causes sand streaking. The forms must be braced and strong enough to hold the concrete. If the forms are to be used again, they must be easily removed and re-erected without damage.

Forms for concrete work are generally constructed from earth, metal, or wood.

- *Earth forms* are used in subsurface construction where the soil is stable enough to retain the desired shape of the structure. This type of form requires less excavation, and there is better settling resistance. Earth forms are generally used for footings and foundations.
- *Metal forms* are used where added strength is required or where the con-

struction will be duplicated at another location. Metal forms are seldom used on home-type projects because of their cost.
- *Wooden forms* are the most common type used in building construction. They are economical, easy to handle and produce, and adaptable to many shapes.

Lumber used for formwork should be straight, strong, and only partially seasoned. Kiln-dried timber might swell when soaked with water from the concrete. If the boards are tight jointed, swelling causes bulging and distortion. Softwoods such as pine, fir, and spruce are light, easy to work with, and readily available.

Remember that the concrete will take on the surface pattern of the formwork, so any lumber that comes in contact with the concrete should have a smooth surface unless a deliberate woodgrain effect is wanted. If a supersmooth finish is desired, use plastic sheets or heavy kraft paper to line the forms.

Wooden formwork requires only basic carpentry skills. The emphasis should always be on strength and stability. Strong, simple butt joints should be used at all formwork corners and seams. Stake and nail the forms firmly in place. Because most formwork is temporary, it's a

Formwork uses basic carpentry principles, such as simple nailed butt joints at corners and joints.

good idea to use duplex (double-headed) nails. Drive them in only to the first head to hold the lumber together, and leave the second head exposed for easy removal.

Release Agents

A *releasing agent* is often applied to the surfaces of forms that contact the fresh concrete. It ensures that the forms will pull away from the set concrete more easily with less chance of damage to the concrete. Commercial form releases are available that are applied with spraying devices. But for most homeowner projects, clean engine oil works just as well. Use a paintbrush or rag to spread a thin coating of oil over the inside of the forms just prior to placing the concrete.

FORMWORK FOR SLABS

Most formwork for slab-type projects is constructed on 2× framing lumber set on edge. Regardless of the slope of the land or drainage pitch required, formwork lumber for slabs must be positioned at true vertical.

The choice of 2×4, 2×6, or 2×8 lumber is determined by the required slab thickness. Keep in mind that the actual dimensions of the lumber are slightly smaller than its nominal dimensions. A 2×4 actually measures 1-1/2″×3-1/2″, a 2×6 is 1-1/2″×5-1/2″, and a 2×8 is 1-1/2″×7-1/4″.

So if 2×4 forms are set directly on the leveled subgrade, the slab thickness would be 3-1/2″. The loss of this 1/2″ of height might not affect the success of the project in many cases, but the problem is easily solved. Simply raise the formwork 1/2″ off the subgrade when it is nailed to the securing stakes. This will leave a small gap at the bottom of the form that must be backfilled with soil prior to placing the concrete. Pack the soil tightly against the outside of the forms to close the gap and prevent leakage.

Staking

Slab formwork is always staked to the ground. Stakes should be made of 1×2 or larger lumber that is strong and free of cracks or defects. Hold the stake tightly against the outside of the form and drive it into the ground at least 6″. Drive the top of the stake flush with or below the top of the form, or cut off the top of the stake flush once it is in position. Stakes that protrude

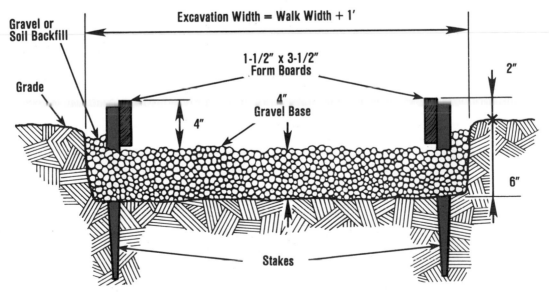

For slab formwork, lumber set on edge can be positioned at a height above the subgrade equal to its nominal dimension. The 1/2″ gap at the base can be backfilled with stone or soil to contain the wet concrete.

All slab formwork is firmly staked in position. Set stakes flush with the top of the form.

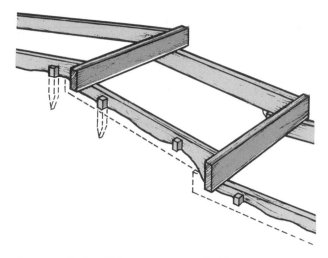

Formwork detail for creating a shallow step in slab formwork

above the top of the form will prevent easy strike-off or screeding of the concrete.

Secure the stake to the form by nailing through the stake from the outside. Stakes should be located every 3' to 4' along the form perimeter, at all corners, and wherever the ends of two pieces of formwork lumber meet in a butt joint.

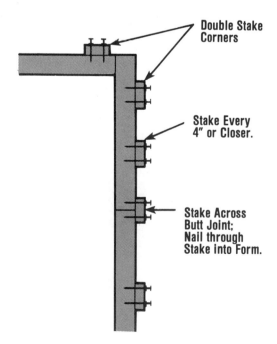

Double Stake Corners

Stake Every 4" or Closer.

Stake Across Butt Joint; Nail through Stake into Form.

Details for securing stakes to formwork

Stepped Forms

The pitch of a walkway running down a moderate slope can be reduced by constructing the walk in a series of shallow steps. If a granular base material is used to prepare the subgrade, stop it 6" to 8" behind the stop board. This ensures that the concrete will be extra thick at the step locations. Be sure to account for this when estimating materials.

Curved Forms

Curved walkways, drives, and free-form patios create pleasing visual effects and blend well with sloping or rolling landscapes. Curved formwork for such projects can be built in a number of ways.

Long strips of 1/4" plywood can be easily bent to form short radius curves. Make certain that the grain of the plywood is running vertically as shown when you size and cut the pieces. Other materials such as hardboard, stiff fiberglass sheeting, or sheet metal can be used to form sharper turns. Whenever the form material is thin, be sure to support it with additional bracing.

To form long radius curves and gentle turns, 1×4 lumber can be used. Drive support stakes deep into the ground because the form will resist

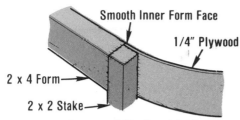

Smooth Inner Form Face

1/4" Plywood

2 x 4 Form

2 x 2 Stake

**Suggested Detail at Joint between
Straight and Curved Forms.**

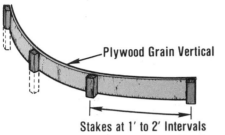

Plywood Grain Vertical

**Stakes at 1' to 2' Intervals
Use 1/4" Plywood or Hardboard for
Short-Radius Curves.**

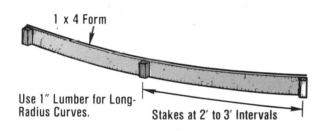

1 x 4 Form

Use 1" Lumber for Long-
Radius Curves.

Stakes at 2' to 3' Intervals

bending. Stakes temporarily positioned against the inside of the form can act as flex points to help bend the lumber and hold it in position until the outer stakes are firmly secured.

To bend thicker material that's not flexible, you can also use the kerfing method. The aim here is to make parallel cuts with a saw across the grain of the wood, always keeping the cuts no deeper than 3/4 of the thickness of the stock. The cuts serve to remove enough of the wood so that it can be bent back on itself. How many cuts you make and how closely they're spaced depends on the amount of bend required. For a sharper bend, make more cuts and space them closer at the bends. It's very important to make all cuts on the outside of the forms.

One tip to keep in mind is that slightly green or wet lumber bends more easily than dry seasoned stock.

You can set the radius of curved corners on larger patios and slabs by using a string tied to a stake set back from the corner as shown. The stake is set at the center of an imaginary circle, and the string is rotated to mark off a section of the circle's circumference that corresponds to the perimeter of the curved corner.

Gentler curves can be set by eye and adjusted as the formwork progresses. Laying out lengths of garden hose or rope can be helpful in visualizing curves and turns. On curved walk-

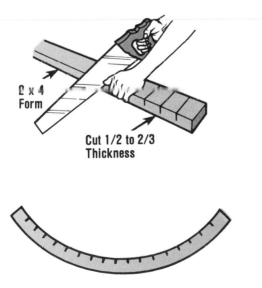

2 x 4
Form

Cut 1/2 to 2/3
Thickness

Formwork detail for constructing curved slab form-work

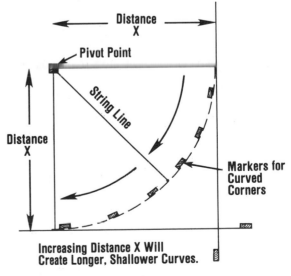

Distance
X

Pivot Point

String Line

Distance
X

Markers for
Curved
Corners

**Increasing Distance X Will
Create Longer, Shallower Curves.**

Method of laying out curved corners

47

ways, be sure that the distance between the forms remains constant. The easiest way to do this is to set support stakes in pairs directly across from one another.

Permanent Forms

In some instances you might want to leave side forms and divider strips in place permanently, either for decorative reasons or to serve as control joints. Control joints, also known as *contraction joints*, induce cracking at selected locations where it will be less conspicuous. Permanent forms are usually made from 1×4 or 2×4 redwood, cedar, cypress, or treated wood. Treated wood needs no special preparation, but the others should be primed with clear wood sealer first. You might also want to cover the top surfaces with masking tape to protect them and preserve their color during the placing and finishing of the concrete.

Use miter joints at the corners and join intersecting strips with neat butt joints. To anchor the outside forms to the concrete after it has been placed, drive 16d galvanized nails at 16" intervals horizontally through the forms at mid-height. Space the nails in a similar fashion for interior divider strips, but drive them from alternate sides of the board.

Permanent formwork of rot-resistant or treated lumber. Use masking tape to protect exposed edges during concrete placement and finishing.

All nail heads must be driven flush with the forms. Never drive nails through the tops of permanent forms. Any stakes that will remain in place for good should be either driven beneath the surface or cut off 2" below the concrete; this way, they won't become a safety hazard.

Elaborate Forms

Formwork for walls, steps, foundations, and molded concrete projects can be considerably more involved than that used in basic slab construction. Forms for walls are commonly constructed of 5/8" or 3/4" plywood sheathing. When properly reinforced with studs, spacers, and ties, plywood provides a smooth finish to the concrete surface. In many cases, accurately constructing the form becomes the main task of the project. Filling the form and finishing the exposed concrete surfaces are almost secondary. For this reason, the formwork details of specific projects are presented in chapters 5, 7, and 8.

Rebar and Wire Mesh

As discussed in chapter 1, concrete expands and contracts with changes in moisture and temperature. It is also subject to heaving and movement from the ground beneath it. Both types of movement can cause cracks to form, particularly in slab work.

You can't prevent these cracks by using reinforcement embedded in the concrete, but you can help distribute them evenly and keep them relatively small. Reinforcement holds cracks closed tightly and prevents them from growing.

The two most common reinforcement materials are iron rebar and wire mesh. Both must be clean and free of rust when used. The mesh or bar should be suspended in the concrete by placing it on flat stones or small pieces of brick. Try to position the reinforcement as close to the middle of the concrete mass as possible.

Rebar is a steel reinforcing bar that comes in standard diameters. For the average job, 3/8" or

Use of rebar reinforcement

Installing wire mesh reinforcement

1/2" diameters are fine. Rebar can be bent to form curves and corners and can be easily cut with a hacksaw. When using several pieces of rebar, overlap adjacent bars approximately 12".

Although rebar is used in both slab and wall work, wire mesh is used most often in slabs. Whenever possible, place it in whole pieces. If you must combine pieces, use a minimum overlap of 6" or the width of one of the openings in the mesh.

Wire mesh is purchased in rolls, so it will have a tendency to curl. Walk on the mesh to flatten it before using it. Cut the mesh with a pair of heavy-duty wire snips. Always wear eye protection during this job, because the strands can snap back unexpectedly.

Finally, when placing mesh or rebar inside the forms, keep the edges an inch or so inside the form. This will ensure that the wire or bar will not be exposed through the finished concrete, a problem that could cause rust streaking.

Chapter 5

CONCRETE WALKWAYS, DRIVEWAYS, AND PATIOS

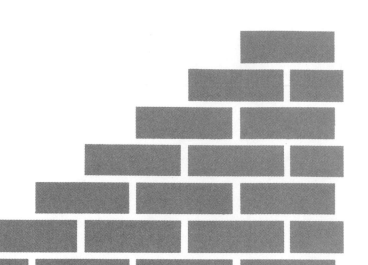

Most concrete projects tackled by average homeowners involve some sort of slab work. Concrete is placed in shallow forms to pave a given area, such as a walkway, driveway, or patio. With the exception of certain design details, all slab work projects are similar in many respects. Most important, slab work requires good surface finishing. Once you understand the basics of screeding, tamping, floating, edging, jointing, and troweling, you'll be able to handle all types of concrete work.

JOINTS IN SLAB WORK

As explained in chapter 1, slight volume changes in concrete and the stresses they produce are unavoidable. To help control stress buildup and the cracking it causes, three types of joints are used in concrete work. They are:

1. Control joints (also called *contraction joints*), which are used to induce cracking at preselected locations

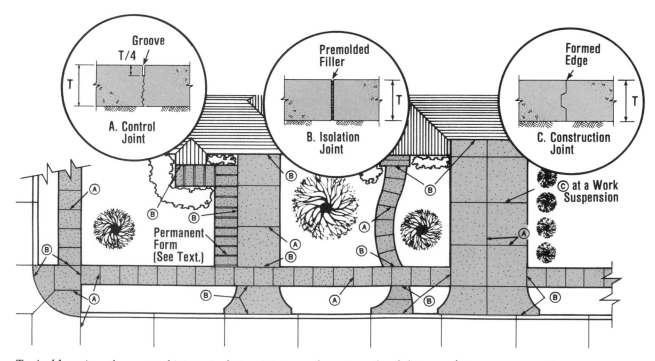

Typical locations for control joints, isolation joints, and construction joints used in concrete work

2. Isolation joints (also called *expansion joints*), which are used to allow movement where the faces of slab abut existing structures
3. Construction joints, which provide stopping places during construction

Control Joints

Control joints penetrate 1/5 to 1/4 the total slab thickness. A joint of this depth provides a weakened section that induces the crack to occur beneath the groove and along that line.

Control joints can be hand tooled, sawed, or formed with premolded or wood divider strips. In driveways and sidewalks, control joints should be spaced at intervals about equal to the slab width. Intervals should not exceed those listed in table 5–1 on page 67. Drives and walks wider than about 10′ to 12′ should have a longitudinal control joint down the center. A large patio might require several longitudinal control joints in its design.

If possible, the panels formed by control joints in walkways, driveways, and patios should be approximately square. Panels with excessive length-to-width ratios of more than 1-1/2 to 1 are prone to crack near mid-length. As a general rule, the smaller the panel, the less likely for random cracking to occur. All control joints

should be placed as continuous lines, not as staggered or offset lines.

Isolation Joints

Most walkways, driveways, and patios are designed to butt against another concrete or masonry structure. For example, a walk may butt against the foundation of your home or garage. It may also butt against the side of a concrete driveway, a set of concrete steps, or a raised street-side curb. Wherever two separate masses of concrete meet, expansion forces also meet and push against one another. And if these forces are strong enough, one of the concrete structures, usually the smaller one, will crack.

To help absorb and negate these forces, an isolation joint should be installed at the joint between the new and existing concrete structures. The isolation joint is sometimes called an *expansion joint*. It consists of a resilient, bituminous fiber strip made especially for this purpose. The strip expands and contracts with the sur-

On larger projects, control joints can be cut into the slab using a power saw equipped with a diamond cut or abrasive concrete blade.

Fiber strip isolation joint nailed to inside of walk forms. When the forms are removed, a concrete curb will be placed adjacent to the walk, with the isolation joint between them. The nails tie the strip to the walk.

rounding concrete and maintains a water tight seal at the joint.

The fiber strips are 1/4" to 1/2" thick and 4" to 6" wide. When the strip is not exposed to foot traffic, such as a joint between a patio and the existing foundation, it can be installed flush with the new slab. However, if the joint is subject to traffic, such as the joint between a driveway and sidewalk, it should be set 1/4" below the surface of the concrete. The strip material is easy to cut with a sharp knife and can be positioned and held in place with stones or several globs of concrete just prior to the pour.

Isolation or expansion joints do not have to be placed at regular intervals in new walkways, driveways, and patios as long as proper control joint spacing is used.

Construction Joints

Sometimes it will be impossible to complete all placement and finishing in a single day. One example would be mixing concrete from scratch for a long walkway. In this case, the job must be broken into manageable sections, with definite stopping and start-up points.

Construction joints are inserted where concrete placement is suspended for thirty minutes or more or along the perimeter of the day's work. Construction joints normally are located and built so as to act as control joints or isolation joints.

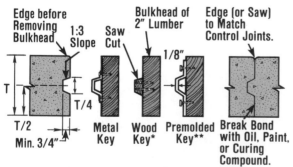

*Beveled 1 x 2 Strip Is Adequate for 4" to 6" Slabs.
**May Be Left in Slab Permanently by Tacking Lightly to Bulkhead.

Details for constructing a tongue-and-groove construction joint for tieing concrete slabs together

To form a construction joint, a bulkhead of 2"-thick lumber is positioned across the formwork. The concrete is placed to this point, with the edge against the bulkhead finished as one side of a control joint. When work is continued, the bulkhead is removed and the adjacent slab placed. The edge that butts the previous day's work is finished as a control joint.

When the work is set on level, stable ground, a simple butt joint is satisfactory for 4"-thick slabs. For thicker slabs or work on more unstable bases, a tongue-and-groove joint should be used to provide load transfer across the joint and to ensure that adjoining slabs will remain level. The tongue and groove is formed by fastening a metal, wood, or premolded key to the face of the wood bulkhead. Be sure the fresh concrete fills the keyed area in the existing concrete, and finish the joint as a control joint.

CONCRETE WALKWAY

If you're a novice concrete worker, you'll find that building a concrete walkway is an ideal introduction to working with this building material. Concrete walks are well within the skills of most do-it-yourselfers, and the scale of the job is much less than a project such as a concrete driveway or large patio.

Planning and Formwork

Planning, site preparation, and formwork construction have been covered in chapters 2 and 4, so from now on, we'll point out only specifics that pertain to the project at hand.

Walkways can be as wide or as narrow as you desire as long as they do not conflict with local building codes and standards. Most municipal sidewalks that front properties are 5' wide, and those running to front entryways are 3' to 4' across. Service walks should be a minimum of 2' wide.

Most walkways are 4" thick, and it's not recommended that you design walks any thinner than the 3-1/2" thickness created by 2×4 lumber set on edge. The only time thicknesses greater

Check formwork depth in several locations to ensure an even pour that does not waste concrete.

Dampen the subgrade prior to placing concrete.

than 4" are required is when the walk abuts a driveway. Thickness in these sections should match your driveway's thickness—5" to 6" if any vehicle heavier than a passenger car passes over it. When sections of the walkway vary in thickness, install isolation joints at these points.

Double check the formwork prior to mixing concrete or accepting ready-mix delivery. All forms should be securely staked and nailed in position and any openings at the bottom of the forms should be backfilled with soil. Forms for walkways must have a slight pitch or slope to one side, so that rainwater will not form standing puddles. When the walk runs parallel to the house or garage, slope the walk away from the structure. On small jobs, you can check for proper pitch using a mason's level as shown here. If used, a 4" to 6" base of 3/4" to 1-1/2" crushed stone or gravel should be level and in position.

All reinforcing rebar or mesh should be properly positioned, and fiber strips for needed isolation joints should be precut and set at their locations. Form release agents should be applied.

The final preparation step for all concrete work is to dampen the subgrade with water. This prevents the ground from sucking water out of the concrete that is needed for proper hydration. But don't overdo it. There should be no standing puddles or muddy areas.

Spreading and Compacting

Place concrete uniformly to the full depth of the forms and as near as possible to the final position. Start placing in one corner and avoid dragging or flowing the concrete excessively. Overworking the mix during spreading causes

Checking for proper pitch away from the building. Notice that when held level, the right end of the level is about 1/2" above the outside form, indicating a slight slope.

Position the concrete without overworking it.

Screed concrete with a straightedge.

an excess of water and fine sand to be brought to the surface.

Pack the concrete tightly against the sides of the forms to eliminate voids and a possible honeycomb finish when the forms are removed. A short-handle shovel or wide hoe should be used to move the concrete. Avoid rakes. They tend to separate the fine and coarse aggregate in the mix.

Once concrete has been spread and compacted to fill the forms, strikeoff and bull floating or darbying follow immediately. It is of utmost importance that these operations are performed before bleed water has an opportunity to collect on the surface.

Concrete should not be spread over too large an area before strikeoff. Nor should a large area be struck off and allowed to set for any length of time before floating. Remember, any operation performed on the surface of a concrete slab while bleed water is present will cause serious dusting or scaling. This point cannot be overemphasized. It is the basic rule for successful finishing of concrete slab work.

Strikeoff or Screeding

Strikeoff is the operation of removing concrete in excess of the amount needed to fill the forms and bring the surface to grade. The operation is usually called *screeding* by experienced concrete workers.

Although special metal tools are available for screeding, a straight piece of 1×3 or 2×4 lumber works just as well. For small jobs, such as the one shown here, screeding can be a one-person operation. For larger areas, enlist the aid of a helper.

Concrete is screeded by moving the straightedge back and forth with a saw-like motion. A small amount of concrete should be kept ahead of the straightedge to fill in low spots. Screeding is done in two passes. Break the work done into 3' sections. On the first pass, as the straightedge is pulled forward, it should be tilted in the direction of travel to create a slight cutting edge. A second pass should be made to remove any bumps or low spots. During the second pass, the straightedge should be tilted in the opposite direction.

Tamping

When needed, the concrete is tamped or compacted after screeding. Tamping helps press larger aggregate below the surface and compacts the mix against formwork and the subgrade.

Tamping can be performed with the flat end of an iron rake, a simple homemade tamper, or a rented commercial tamper. The first two choices are best for walks and small slabs.

Simply hold the rake or tamper in a vertical position and push down the end into the wet

If the concrete contains large, coarse aggregate, tamping with a rake or tamping tools may be required.

On smaller jobs, first floating can be performed with a wooden or metal hand float.

Wooden darbies can also be used to level and smooth concrete after screeding.

concrete for a fraction of an inch—just enough to force the top stones beneath the surface. If the coarse aggregate is small and the subgrade firm and well-prepared, the need for tamping might be minimal.

Floating or Darbying

Immediately after screeding and tamping, the concrete must be floated using a wooden or metal float or darby. Like tamping, floating pushes larger pieces of aggregate below the surface. It also removes any screeding or tamping marks or ridges and helps bleed water to rise to the surface.

This work can be done using a hand-held float or darby or a larger bull float, which is operated back and forth over the surface like a push broom. The finishing surface of these tools can be made of wood, rubber, or a light metal, such as magnesium.

Hand floats are best for small jobs and tight spots. Darbies and bull floats can cover large areas quickly. Regardless of the tool used, do not overwork the concrete at this point in the finishing process. One, or at the most two, passes will smooth out imperfections and allow the bleed water to rise to the surface.

When using a hand float as shown, swing the tool in wide circles, holding up the leading edge of the tool slightly to keep it from digging into the concrete. Pick out any large aggregate that cannot be easily pushed beneath the surface.

Darbies are used like a combination screed and float. Begin by holding the darby flat against the surface of the concrete and work it from right to left with a sawing motion, cutting off bumps and filling depressions. When the surface is level, tilt the darby up slightly and move it across the surface to fill any small holes left by the sawing motion.

When a bull float is used, it should be pushed forward with the leading edge of the float raised so it will not dig into the surface of

Push the bull float across the surface with the front edge raised slightly; pull it back with the blade positioned flat on the surface.

the concrete. The tool should be pulled back with the float blade flat on the surface to cut off bumps and fill holes. If holes or depressions remain and no excess concrete is left on the slab, additional concrete should be shoveled from a wheelbarrow and the surface bull floated a second time.

Immediately after floating or darbying, cut the concrete away from the forms to a depth of 1" using a pointed mason's trowel. Separating the top edge of the slab from the formwork will make edging easier later in the finishing process.

Finishing Concrete

The final finishing operations must wait until the bleed water brought to the surface by first floating or darbying has left the surface and the concrete stiffens slightly. This waiting time, which is absolutely essential to obtain durable surfaces, varies with the wind, temperature, and relative humidity of the atmosphere. It is also affected by the type and temperature of the concrete. On hot, dry, windy days, this waiting period can be very short. On cool, humid days it can be as long as several hours.

When air-entrained concrete is used, there might be very little or no waiting. Begin finishing when the water sheen is gone and the concrete can sustain foot pressure with only about 1/4" indentation. A water sheen might not be visible on some air-entrained concretes.

Edging

Edging is the first finishing operation. It is done to produce a smooth, rounded edge at all corners. A rounded edge helps prevent chipping or damage, especially when the formwork is removed. Edging also compacts and hardens the concrete near the forms, an area where floats and trowels are less effective.

Stainless steel edgers with a 1/2" radius are recommended for walkways, driveways, and pa-

Cut the top 1" of concrete away from the forms using a mason's trowel. This makes edging easier.

Edging forms a slight curve on all corners to help prevent chipping.

tios. Before beginning, be sure the concrete has set up or stiffened sufficiently to hold the shape of the edger. The edger should be held flat on the concrete surface. Slightly tilt up the front of the edger when moving the tool forward. When moving the tool back over the edge, tilt the rear slightly. Be careful to prevent the edger from leaving too deep an impression, because these indentations might be difficult to remove later. In some cases, edging is required after each finishing operation. Final edging is done after final troweling.

Jointing

Immediately after edging, control joints can be pressed into the concrete. Control joints in walkways are usually formed with a jointing tool. Use a straight board to guide the tool as shown. Rest the board on the side forms, perpendicular to the edges of the walk. Hold the jointing tool against the board as you move it across the slab.

To start the joint, push the jointing tool into the concrete and move it forward, applying pressure to the back of the tool. After the joint is cut, turn the tool around and pull it back over the groove to give the joint a smooth finish.

Edging and jointing must be done carefully. If the surface is gouged out by hand edgers or jointing tools, the damage might be difficult to repair. If the tool kicks up a large piece of aggre-

gate, remove it, fill the spot with a little concrete containing only sand, and retool the spot.

In walkways, control joints should be located at intervals equaling no more than 1-1/2 times the walk's width. So for a 3'-wide walk, control joints should be located every 4-1/2' or less. For wider walks, spacing should equal walk width.

Cutting Joints. Control joints can also be cut into the concrete using a circular saw equipped with a masonry cutting blade. Sawing should be done when the concrete surface is firm enough not to be torn or damaged by the blade. This is usually 4 to 12 hours after placement. A slight raveling of the sawed edges is normal and indicates proper timing. If sawing is delayed too long after placement, the concrete might crack before it is sawed, or cracks might develop ahead of the saw. Cut the joint in several passes, lowering the blade slightly each time.

Wood Control Joints. The final method of forming a control joint is to embed a strip of 2×4 redwood, cedar, or cypress into the concrete at the joint locations.

Second Floating

Following edging and jointing, the surface is floated once again. Floating will embed any large aggregate that has risen to the surface, remove

Cutting control joints using a jointing tool. A straight-edge ensures straight lines and a steady hand.

A second floating produces a slightly rough, textured finish. This job uses permanent divider strips to produce an interesting patterned effect.

any imperfections left by edging and jointing, and compact the concrete at the surface.

Floating creates a rough concrete finish. Rough concrete provides good traction, so it is particularly suited for walkways, driveways, pool decks, and other areas frequently exposed to water. It is also more durable than smoother concrete finishes.

Magnesium floats are light, strong, and glide easily over concrete surfaces. They are recommended for most work, especially for air-entrained concrete. Wood floats generate a little more drag, but they produce a slightly rougher finish.

The hand float should be held flat on the concrete surface and moved with a slight sawing motion in a sweeping arc to fill in holes, cut off lumps, and smooth edges. Stop floating when the entire slab is fairly smooth and the float leaves no visible marks. At this point there is nothing to be gained by continued floating. For denser finishes, you must switch to a trowel.

For larger jobs, second floating can be done using a bull float or darby. Touch-up with a hand float is usually required.

On some walkways, the marks left by edgers and jointers are used as decorative borders to break up the flatness of the slab. As these marks are removed by floating, you should re-edge and re-joint if you desire this treatment.

Troweling

When a coarse finish is desired, edging, jointing, and floating might be the only steps required. But when a denser, smooth finish is desired, troweling is necessary. One to three trowelings are possible, with each pass generating a smoother finish.

Troweling is done after the second floating. It should never take place on concrete that has not been floated a second time. Troweling after screeding and first floating/darbying will not produce good results.

Make sure all water sheen is gone before beginning troweling. Delaying troweling too long results in concrete that is too hard to work, but

Troweling produces a hard, smooth finish.

the tendency among less experienced finishers is to begin troweling too soon.

For the first troweling, the trowel blade should be held flat on the surface. Tilting the blade produces ripples that are difficult to remove without tearing the surface and causing popouts. The concrete should feel like wet, packed sand beneath the blade. Move the hand trowel in a sweeping motion, with each pass overlapping one-half of the previous pass. Using this technique trowels all areas two times. The first troweling should produce a surface free of defects, which is about all you would want for a walkway. Dense, smooth finishes become very slick when wet.

If the slab is protected from the weather or other factors outweigh the need for traction, you might want to trowel a second or third time. The smooth, almost glossy finish that multiple trowelings produce is easy to keep clean and is ideal for decorative inscriptions and stampings.

There should be a lapse of time between the first and second trowelings to allow the concrete to become harder. Begin the second troweling when pressing against the surface with your hand produces only a slight indentation.

For the second troweling, tilt up the blade of the tool slightly and increase the finishing pres-

CONCRETE

Steps in producing exposed aggregate finishes: spreading stone, pressing beneath surface, and removing excess mortar (top, left to right). Decorative aggregate walk with permanent divider strips (above). The aggregate in this concrete driveway adds color and texture (right).

CONCRETE

Plenty of help is important for larger projects, such as this concrete driveway (above). This basic concrete driveway is designed with a uniform slope away from the garage (above right). Troweling the surface of dry-shake pigmented concrete (right).

Stamping colored concrete to create a simulated brick paver pattern.

All three patterns shown above were created by stamping or grooving flat, colored concrete.

This concrete walk features brick border paving and decorative wood inlays.

Deep cut grooves control cracking and create visual interest in this basic concrete patio.

This landscaping scheme smoothly combines concrete walkways, steps, and walls.

CONCRETE

The wide landings of this stepped entrance stress the smooth, straight lines of basic concrete work. Built-in flower beds add color accents.

This terraced entrance walk alternates between white and decorative aggregate concrete. Slump block is used for the privacy wall, as well as the landscape edging.

These travertine finished steps feature wide overhanging treads. This is strictly a warm climate decorative finish.

This circular free-form patio combines colored concrete with a stamped pattern.

BRICKWORK

Brick is the ultimate all-purpose building and paving material.

Mortarless brick paving leading to a mortared brick stoop.

This gently curved brick wall features special rounded capping brick.

BRICKWORK

This brick screen wall uses brick columns for support and visual variety.

Mortared brick paving for patio and step combination (above). Slump-faced block combine with stretches of open screen block to create an appealing privacy wall (above right). Interlocking concrete pavers create interesting geometric patterns (right).

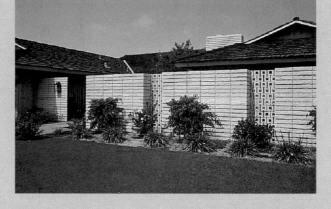

Many waterproofing coatings are as simple to apply as paint.

Applying surface-bonding cements to dry-stacked block. The work is similar to applying a finish stucco coat or plaster parge coat.

Brick patio and planter combination

A corbelled mailbox post, raised tree well, and mortarless barbecue are just several samples of easy, practical brick projects.

STONE

Stone veneer adds variety to many residential settings. Both the chimney and entranceway shown above are veneered with lightweight manufactured stone that is easy to install.

Stone retaining wall topped with flagstone capping

This mortarless stone wall dates back to the early 1800s.

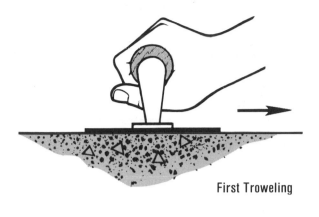

First Troweling

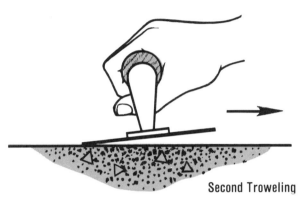

Second Troweling

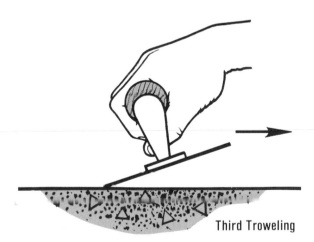

Third Troweling

Trowel positions for first, second, and third trowelings. Increase downward pressure while executing each troweling.

sure. If the desired finish is not obtained with this troweling, there can be a third troweling after another lapse of time. The final pass should make a ringing sound as the tilted blade moves over the hardening surface.

Curing

As mentioned in chapter 1, curing is one of the most important steps in concrete construction and, regrettably, one of the most neglected. The purpose of curing is to keep the concrete moist so hydration can continue uninterrupted for five to seven days after placement. This greatly increases the concrete's strength and durability.

Once you begin a concrete project, never allow the surface of the concrete to dry out completely. This means you might have to mist the surface with water during the final finishing steps. And always begin formal curing immediately after final finishing.

Moist curing can be done in a number of ways. The best method is to hose down the slab at regular intervals or to set up sprinkling equipment that provides a continuous misting. In all but the hottest, driest climates, a good dousing after nightfall will keep the surface wet until morning.

Covering the concrete with wet burlap, sheets, or blankets will slow down but not stop the evaporation process. One way to seal in the water at the surface is to cover the slab with plastic sheeting. Plastic sheeting must be laid flat and completely sealed at joints and along edges. If the sheeting becomes wrinkled, patchy discoloration of the concrete can occur.

Covering with plastic sheeting after applying a water spray will help retain surface moisture during curing

Another way to seal the surface against moisture loss is to apply a pigmented curing compound. These compounds are applied by spraying soon after the final finishing operation. The surface should be damp but not wet with standing water. Complete coverage is essential. Applying a second coat at right angles to the first is recommended if there is any doubt as to the thoroughness of the coverage.

Do not apply curing compounds in late autumn in climates where deicers will be used in the coming months. The compound may not give the concrete sufficient time to air dry. Complete air drying is needed to increase the concrete's resistance to scaling caused by deicers.

Curing compounds are most often used by contractors who enjoy the convenience of a one-time application. But for homeowners doing their own work, hosing and sprinkling methods are preferred.

Removing Formwork

With the exception of bulkhead boards used to create construction joints, leave all formwork in position until after the initial curing period of five to seven days. The urge to remove the formwork will be great, but you are only tempting fate. The chances of chipping edges or marring finishes are greater early in the curing process.

When you do remove the form boards, avoid applying pressure to the concrete. Remove the nails securing the stakes to the forms and pull the stakes straight up and out of the ground. Pull the form away from the concrete with the tips of your fingers. Never pry boards away with a bar or large screwdriver inserted between the form and the concrete. Several light raps with a hammer on the outside of the form can also break the bond between the lumber and concrete. If oil or a release agent was applied, this should not pose a problem.

Once the forms are removed, rub off any rough spots with a pumice stone or ordinary brick. If there are any noticeable hollows or honeycombed areas, fill them in with a mixture of one part cement to two parts fine sand and water.

This driveway clearly shows the proper use of side-to-side pitch and downward slope. It uses an exposed aggregate finish with brick paving strips as control joints.

DRIVEWAY

Constructing a driveway is similar to constructing a walkway on a very large scale. A wide, shallow form is built, filled with concrete, then screeded and finished. Ready-mix concrete is normally used on a job of this size, and excavation and site preparation can be extensive. But the personal satisfaction and financial savings of seeing a larger project through to completion are great.

Building Codes

Driveways almost always cross or abut a public way, so all municipalities are very strict concerning their design and construction. This is one project where working hand in hand with the local building authorities is an absolute must. Building authorities are concerned about the overall slope of the driveway, its profile, or side view as it crosses public walkways, and the manner in which it abuts curbs and public streets. If a curb and gutter have not been installed, it is advisable to end the driveway temporarily at the public sidewalk or property line.

An entry of gravel or crushed stone can be used until curbs and gutter are built. At that time, the drive entrance can be completed to meet local requirements.

If the driveway is built before the public walk, it should meet the proposed sidewalk grade and drop to meet the gutter (if no curb is planned) or the top of any low curb.

While in the planning stages, consideration should be given to other elements that can make a driveway a beautiful approach to your home rather than a plain path to your garage. Consider offstreet parking, turnaround areas for safe head-on street entry, paved areas for recreation and so forth.

Specifications of commonly accepted driveway designs will be outlined, but check all plans with local authorities. Don't imitate other driveways in your neighborhood. They might be outdated. And always have the building department check the site after you have excavated and set up forms but before you place any concrete.

Determining Slope

Slope is simply the driveway's vertical drop per foot of length. A driveway that drops 1' for every 10' it travels has a 10% slope (1' ÷ 10' = 0.10 or 10%).

To measure the slope of the site, drive in a long stake perpendicular to the ground at the base of the driveway. Attach a string and line level to this lower stake and run the string out perpendicular from the lower stake up the slope a distance of 10', 20', or 30'. Working with round figures such as these will make calculations easier. Drive another stake into the ground at your selected distance. Tie the string to this upper stake at ground level. Now readjust the location

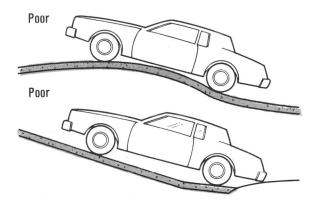

Poor

Poor

Poor planning can lead to clearance problems at the point the driveway meets the sidewalk.

of the line on the lower stake so the line level again reads level. When this is done, measure the distance between the point the line contacts the lower stake and the ground. This is the vertical drop at ground level from the upper stake to the lower stake.

Divide this distance by your selected run distance to calculate the slope. For example, if the drop is 2.5' over a 20' run, the slope is 2.5' ÷ 20' = 0.125 or 12.5%. Measure the slope over the entire run of the driveway. Slope should be as continuous as possible or, if needed, should build ever so gradually to avoid striking the vehicle's underside.

A slope or grade of 14% (1-3/4" vertical rise per each running foot) is the maximum recommended by most building codes. A 14% slope is substantial. If your site exceeds this limit, professional advice might be in order.

The shape of the drive as it crosses the sidewalk is known as its *profile*. Some codes will insist that it is level with the walk, while others might allow the drive to be a little lower. The slope between the walk and the road is also critical. As shown in the illustration, when the profile is too steep, shorter vehicles tend to hang up in the middle and longer vans and wagons might strike their rear bumpers.

Design Factors

Driveways for single-car garages are normally 10' to 14' wide and must be at least 14' wide

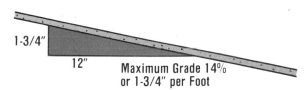

1-3/4"

12"   Maximum Grade 14%
or 1-3/4" per Foot

Slope is calculated by dividing the length of the run by the rise or drop.

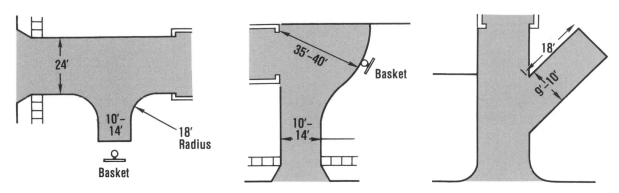

Turnaround and parking strip details

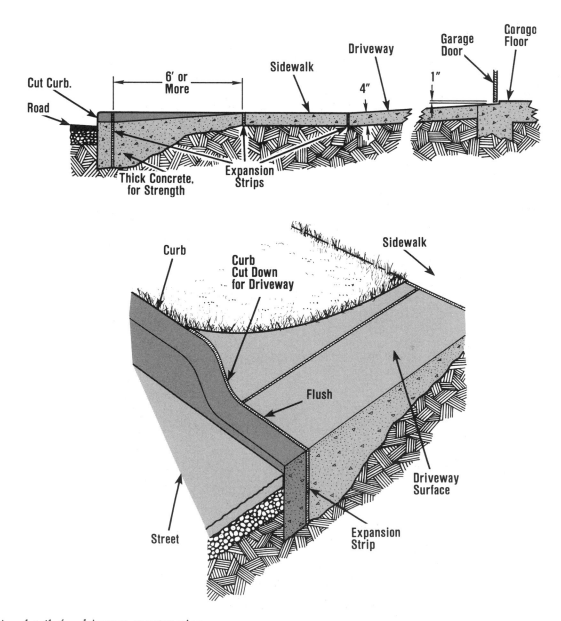

Design details for driveway construction

if the driveway curves. A driveway serving a two-car garage is normally about 24' wide. When a single-lane driveway is built to serve a two-car garage it should begin to widen 15' to 20' before reaching the garage entrance. As shown, driveways can be designed with turnaround areas to eliminate backing out into the street. Parking strips are also an excellent idea.

Thickness. Slab thickness is normally 4" for driveways serving passenger cars. Those serving trucks should be 5" to 6" thick. Consider building up slab thickness near the street entrance so it can safely bear the weight of an occasional delivery truck or a heavy vehicle looking for a convenient turnaround point.

The driveway should be set 1" below the entrance to your garage to prevent water from easily flowing into the building. The opposite end should be flush or a little higher than the top of the curb or road. Setting the driveway 1" to 2" above road level will keep water and dirt off its surface. Like a sidewalk, the top surface of the driveway should be about 2" above grade.

Pitch. Proper pitch is critical for keeping the driveway free of standing water. A side-to-side pitch of 1/4" per foot can result in a significant grade difference across a wide drive. With a 24'-wide slab, the total pitch would be 6". Obviously, setting such a large pitch involves more than simply lifting one side of the form higher than the other as with a walkway. Grading the subgrade is one answer, but even this is difficult to do over such a wide area.

A more practical solution is to build a crowned design. As the illustration shows, the drive is built in two separate sections with an isolation joint down its center. This design splits the overall slope in half. The center of the drive is the highest point, and water now flows off the slab to the sides. Splitting the drive into two major sections is often desirable from a work standpoint as well.

Construction Tips

Site preparation and excavation is similar to that used for walkways. Once excavation is complete, begin placing the formwork at the garage and work outward to the mouth of the driveway. Make certain the top of the form is slightly below garage floor level and that the drive will slope away from the garage door.

On crowned designs, install the center guide board at the proper elevation as you install each pair of side forms. On drives that pitch side to side, installing a temporary screeding board down the center of the form is also recommended. The temporary screed board is set at an elevation that will maintain the proper side-to-side pitch. The board helps break down screeding into manageable sections. Once screeding in that section is complete, the board is pulled from the fresh concrete and voids are filled by hand, then darbied or floated smooth.

As the work proceeds, constantly check for proper slope, pitch center to side or side to side, and formwork depth. Formwork details for a curved treatment at the driveway entrance are shown in the illustration.

Once formwork is complete, spread the granular base. Crushed stone or gravel 3/4" to 1-1/2" in diameter should be used. A notched strikeoff board should be used to set the base at the proper level.

Next, cut and position all expansion strip material. Expansion joints will be needed at the garage foundation, down the center of a crowned driveway, at all sidewalk and curb joints, and against any retaining walls that abut the driveway. Finally, install wire mesh or rebar

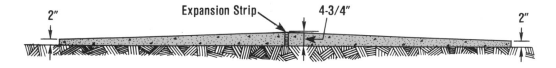

Narrower drives can use a cross slope design, but crowned designs are preferred for wider driveways.

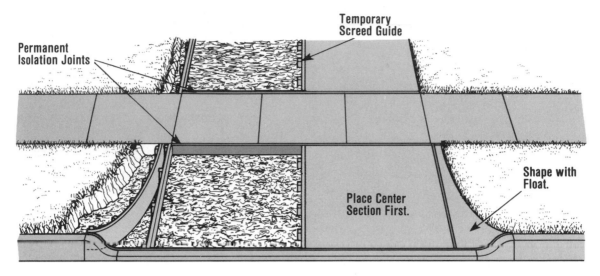

Formwork and finishing details for a curved mouth driveway abutting a cut down curb

reinforcement, if used. The reinforcement should not pass under the expansion joints. Cut it several inches short of the joint. For long drives, you might want to install reinforcement only in the final few yards nearest the entrance. This stretch of concrete will bear the load of turn-around traffic and delivery trucks.

At this point in construction you should have the project reviewed and approved by local building authorities.

Concrete Placement

Review the guidelines given in chapter 3 on working with ready-mix concrete. This certainly qualifies as a large-scale project, so have plenty of eager help on hand. The width of the forms makes it impossible to do all work from the sides, so have the crew wear rubber boots.

Have at least two screed boards long enough to span the forms or reach the temporary screed boards, and construct several hand tampers so the crew can quickly tamp and compact the poured concrete in place. A long-handle bull float is a must for this project, as are several kneel boards.

If the driveway is fairly level, you can begin placement at either end. Pour the full width of the driveway, or pour to the center board of a crowned driveway. Pour on both sides of a temporary screed board. Work in the concrete against the sides of the form. Double check the positioning of all expansion strips.

Once 3' to 4' of the form are filled, begin the screeding process. Don't allow the placing crew to work too far ahead of the screeding team or teams. Once any temporary screed board is no longer needed, pull it from the concrete and fill in the void. Compact the screeded concrete with the tamper and follow with bull floating. On a long pour it is conceivable that placement, screeding, tamping, and bull floating all will be taking place at once on various sections of the slab.

If the site is more steeply sloped, begin placing at the low end of the drive. Fill about 3' to 4' of formwork and begin screeding uphill. If the concrete slumps or runs back downhill, wait several minutes and screed uphill again. Do not resume placement until the screeded section stabilizes. This technique allows you to work uphill with the stiffened section of slab holding the fresher concrete in position. If you must allow the concrete to begin setting up in order to work uphill, don't neglect the finishing process for these sections. Timing of edging, jointing, floating, and troweling operations is critical. You might not be able to wait until one operation is complete for the whole slab before beginning another.

Finishing. Like walkways, most driveways benefit from a coarse, high-traction finish. The steps involved would be screeding, tamping, first bull floating or darbying, and second bull floating after the slab has lost its sheen.

Edging and jointing are done before second bull floating. Edging is not absolutely essential, but makes for a more professional job. Control joints can be hand grooved or cut. Cut joints often make more sense for larger projects. They allow you to concentrate on other finishing jobs during the critical setup period. Control joints are then cut a little later, after the rush of getting the concrete in place has passed.

Power Troweling. Troweling is also possible for a slightly denser finish. Unless the crew includes several experienced masons, there's no way you can reasonably expect to hand trowel a large slab before it hardens beyond workability. The only alternative, other than pouring in smaller sections, is to rent a power trowel. These are usually gas-driven machines with several large, twirling troweling blades. The power trow-

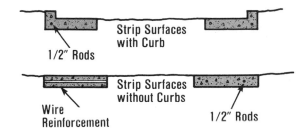

Construction details for concrete strip driveways

el is used after second floating when the surface can bear the weight of the machine and operator without indentation. This is usually a little later than you would begin hand troweling.

Glide the machine back and forth over small areas of the slab until the desired finish is obtained. A *light* sprinkle of water ahead of the machine will reduce friction and lead to a smoother finish.

Strip Driveway

An alternative to the basic slab driveway is the strip driveway. It has one strip of concrete under each wheel, approximately 3' to 4' wide and 5' center to center, with space between the strips. One caution with this type of driveway, however, is that it shifts more easily with changes in ground conditions and should always use reinforcement, preferably iron or steel rebar.

CONCRETE PATIO

A patio is generally placed adjacent to a rear or side door to provide easy access to the interior of the home. However, a patio can also be freestanding or connected to the home by a separate walkway or wing.

Stripped to its bare minimum, a concrete patio is no more than a strategically placed 4"-thick slab. Yet a patio provides one of the finest opportunities for blending unique design and surface treatments with the practicality of concrete construction.

Constructing a patio should be more than paving a section of your backyard. Consider

A gas engine power trowel is the only practical way to impart a smooth, dense finish on larger driveways, patios, and other slabs.

A well-designed patio is both beautiful and functional.

curved or free-form shapes to avoid the monotony of rectangular or square layouts. Employ permanent formwork, exposed aggregate finishes, or colored concrete. Design in permanent flower beds or tree wells, or set aside specific areas for a barbecue pit, dining area, lounging spot, children's sandbox, or recreational area.

You might want to combine concrete, brick, flagstone, and wood construction in any number of interesting and creative combinations, such as using an alternating surface pattern of concrete and brick paving. You can also construct one end of the patio as a slab foundation (see

Many interesting patios begin as simple slab construction. The concrete can be given a decorative surface finish or a mortared overlay of brick or flagstone.

page 96) and build a screen or privacy wall of brick or decorative concrete block.

Location

Orienting your patio in relation to the sun is an important design consideration. Northern, southern, eastern, and western exposures all have their advantages and drawbacks.

It might be helpful to spend some time scouting the proposed site during seasons and times of most anticipated use. Set up a lawn chair and relax. Note the position and intensity of the sun during various times of the day. When will the patio be bathed in sunlight and cooled by shadows of surrounding buildings and trees? From what direction do prevailing winds blow? Are they comforting or annoying? What can you see from your proposed patio? And, more important, who will be able to see you? Does the site lend itself to landscaping with protective trees or screening walls? Is there a tree you want to work around rather than remove from the site? All these considerations are important. The following sections briefly summarize the merits of the possible exposures.

East. A patio facing east will be cool after high noon, because it will be shaded by the house. Eastern exposures are great for hot summer climates, but side screening might be desired to quell the chill of winter.

West. Western exposures offer plenty of sun year-round, but excessive heat and glare can be a problem on summer afternoons and evenings—the precise time most people want to use the patio. Landscaping with trees or screening shrubbery can offer relief and add to the natural setting of the patio. Work around any existing trees or natural screens.

North. A true northern exposure has little to offer, unless the patio is enclosed. The patio will receive minimal sun, which can make it a cold, damp area in all seasons except summer.

South. Southern exposures receive varying degrees of sunlight throughout the year. This will help warm the patio during winter months and make spring and autumn ideal seasons to enjoy outdoor living. A south-facing patio can become quite warm during the height of summer, but glare from the setting sun will be less of a problem than with a western exposure.

Design Factors

The distance from the top of the finished patio surface to the doorway should be no more than 8"—the maximum comfortable distance one can step. If your patio will be more than 8" below a doorway, consider building it up with a thicker subgrade of crushed stone. An alternative is to build a wide concrete step atop the patio, adjacent to the doorway.

The patio must pitch away from the house at a rate of 1/8" per running foot. As mentioned earlier, a 4" slab thickness set 2" above ground level is suitable for most patios. If a 4"-thick granular base of 3/4" to 1-1/2" of crushed stone is not being used, keep the excavated site as level as possible without losing the proper pitch. Excavating deeper than needed will waste concrete when a granular base is not used.

When the patio will abut an existing structure, begin layout, excavation, and form construction where the patio meets the structure.

This ensures that the top of the patio will be at the desired height and slope will be away from the structure. Use the 3-4-5 principle of right triangles to set the patio square with the adjacent structure and double check all dimensions and diagonal measurements. When forms are in place, check for proper pitch away from the building using a level. An expansion joint is needed where the patio abuts an existing structure or wall. Control joints should be spaced as per the guidelines listed in table 5-1. Rebar or wire mesh reinforcement is recommended in climates subject to freeze–thaw cycles.

Openings in the slab for tree wells, flower beds, sandboxes, etc., are made by constructing forms of the proper size and leveling them with the perimeter forms. Stake the forms on the inside and cut the stakes flush with the top edge for easy screeding. When the slab is poured, the forms act as a dam to keep concrete from flowing into the areas.

Slab Thickness (in.)	Slump 4" to 6"		Slump Less than 4"
	Maximum-Size Aggregate Less than 3/4"	Maximum-Size Aggregate 3/4" and Larger	
4	8	10	12
5	10	13	15
6	12	15	18
7	14	18	21
8	16	20	24

TABLE 5-1:   SPACING OF CONTROL JOINTS (IN FEET)

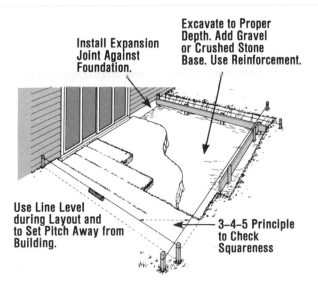

Install Expansion Joint Against Foundation.

Excavate to Proper Depth. Add Gravel or Crushed Stone Base. Use Reinforcement.

Use Line Level during Layout and to Set Pitch Away from Building.

3-4-5 Principle to Check Squareness

Layout and site preparation for a typical patio design

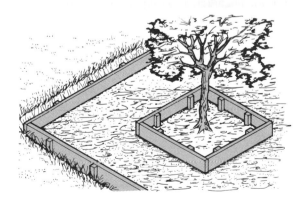

Formwork details for constructing an opening in slab work

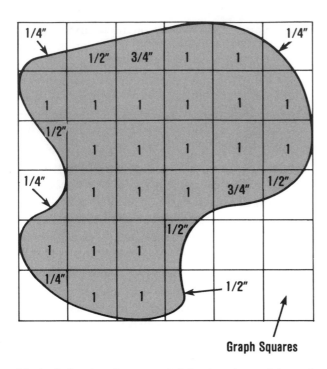

A

B

Method of estimating materials for free-form slabs and patios

Free-Form Designs

Free-form patios can be oval, circular, kidney-shaped, or just about any shape imaginable. Formwork is constructed of thin plywood strips or kerfed 1"-thick lumber.

Keep in mind that while curved or oval shapes might be visually pleasing, they will also reduce the amount of usable patio area. For example, a square patio measuring 20' on a side offers 400 square feet of surface area. A circular patio with a 20' diameter offers about 315 square feet of living space.

Estimating concrete and crushed stone needs can also be a bit tricky. The most accurate method is to draw a proportional outline of the shape on graph paper with each box representing 1 square foot of area. You can then estimate the number of full, 1/4, 1/2, and 3/4 squares and add up the overall area covered.

C

D

Placement and Finishing

Placement is similar to a large driveway project. Temporary screed guides will be needed

Using temporary screed guides to level a large patio: (A) stake the guides at the correct height, (B) screed across guide, (C) remove the guide, and (D) bull float with a homemade wooden bull float.

unless the design requires permanent formwork or is otherwise broken down into more manageable stretches. Be sure the top surfaces of any permanent formwork is fully protected against exposure to fresh concrete.

Following screeding and tamping, the surface of the slab can be finished using any number of different methods. As mentioned earlier, patios are a great opportunity to use more elaborate finishing techniques, such as exposed aggregate, stamped textures, or imitation grooved flagstone. These finishing options are covered in complete detail in the following chapter. If a troweled finish is selected, it should be moderately smooth for easy cleaning, but not too slick.

SHUFFLEBOARD COURTS

Shuffleboard is a great pastime for people of all ages. Official courts measure 6' wide and 52' long, but scaled down courts, as small as 3' wide and 28' long, can be used. The court can be built separately or incorporated as a section of a patio or driveway.

The slab requires a smooth finish and thorough reinforcement to prevent cracking. No control joints are used. The site should be stripped of all sod, down to a firm base. A 4" layer of firmly tamped coarse sand or crushed cinders should be used as a base in soft subgrade.

The overall slab is 5" thick. It is layered in two separate courses. The 3" bottom layer uses a more economical 1:2:3 mix, while the 2" top course uses a smooth finishing 1:1:2 mix with 1/2" maximum coarse aggregate. A layer of wire mesh is placed on top of the lower course, prior to placing the top course. A coating of concrete bonding adhesive will guarantee maximum bonding between the two courses.

The top coat should be given a dense troweled finish and wet cured for seven days. After curing, a coat of clear concrete sealer will decrease surface friction. Paint on 3/4" lines and numbers after the sealer has dried using exterior latex paint formulated for application on concrete. Coloring with concrete stains requires a different method. After masking the lines and

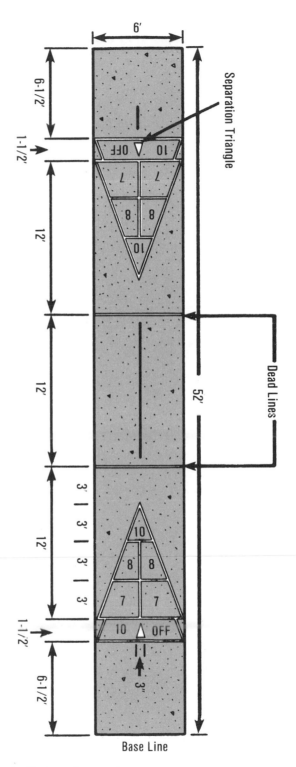

Shuffleboard court dimensions

numbers, apply the base color. Once dry, the masking is reversed and the second color is applied to lines and numbers. The sealing coat or wax is applied seven to ten days later.

SIMPLE SLAB PROJECTS

Small concrete slabs provide strong, level bases for many items inside and outside the home. Here are just a few ideas for Saturday afternoon projects. They're also an effective way to dispose of leftover concrete from a bigger job.

Gallery of slab projects

Chapter 6

SPECIAL CONCRETE FINISHES

There are many alternatives to the standard rough-floated and smooth-troweled concrete finishes described in chapter 5. Adding a unique finish to a rather ordinary slab of concrete can greatly enhance the beauty of the project.

Color may be added to the concrete through the use of white cement and colored pigments or by exposing colorful aggregate at the concrete's top surface. Textured finishes can be as varied as your imagination allows—from a smooth polished appearance to the roughness of a pebble path.

Random and geometric patterns can be scored or stamped into the concrete so it resembles stone, brick, or tile paving. Other interesting patterns can be created by using redwood divider strips to form panels of various sizes and shapes: square, rectangular, or diamond. Special techniques are available to make concrete surfaces more slip resistant. Sparkling finishes are also possible.

COLORING CONCRETE

Colored concrete is an ideal way to brighten a patio or pool deck area. It is also the first step in creating more authentic-looking stamped and embossed patterns for simulated brick, Belgian block, and flagstone paving.

Concrete can be colored in one of three ways:

1. Add pigment to the mix.
2. Apply pigment and work it into the surface of fresh concrete.
3. Apply paint, colored coatings, or pigmented stains to cured concrete.

Pigmenting

With this method, the concrete mix is integrally colored by adding a certain type of mineral oxide pigment specially prepared for use in concrete. Table 6–1 lists the type of mineral oxides used to generate popular colors.

The amount of pigment used should never exceed 10% by weight of the cement used in the

A striking walk of random-size concrete stepping stone with beach pebble exposed aggregate finish

TABLE 6-1: MINERAL PIGMENTS FOR COLORED CONCRETE FLOOR FINISHES

Color Desired	Commercial Name	Approx. Quantities Required, in Pounds, per Sack of Concrete	
		Light Shade	Dark Shade
Grays, blue-black, and black	Germantown lampblack	1/2	1
	Carbon black	1/2	1
	Black oxide of manganese	1	2
	Mineral black	1	2
Blue	Ultramarine blue	5	9
Brownish red to dull brick red	Red oxide of iron	5	9
Bright red to vermillion	Mineral turkey red	5	9
Red sandstone to purplish red	Indian red	5	9
Brown to reddish brown	Metallic brown (oxide)	5	9
Buff, colonial tint, and yellow	Yellow ochre	5	9
	Yellow oxide	2	4
Green	Chromium oxide	5	9
	Greenish blue ultramarine	6	9

mix. White portland cement will generate brighter colors or lighter pastel shades when used with light-colored sand. Use white portland cement for all colors except black and dark grays. All materials in the mix must be carefully controlled *by weight* to maintain a uniform color from batch to batch.

Start with a perfectly clean mixer or mortar box. Tools should also be clean. Always mix the cement and color pigments when dry to ensure thorough mixing before adding any water to the mix. Adding pigment after adding water can result in color streaks of varying intensity. Pigmented concrete can be placed using a one-course or two-course method.

One-Course Placement. This is the simplest method. With the exception of adding pigment during mixing, the job is handled like all others. Placement and finishing techniques are standard. The end result is a mass of concrete with uniform color all the way through the slab. However, since only the top 1" to 2" of most slabs are ever in view, many masons view this technique as a waste of expensive pigmenting agent. For this reason, they prefer the two-course method.

One-course placement is better suited for small slab projects and is always used for walls and other forms where more than one major

surface will be exposed to view. When using one-course placement, make sure the subgrade is uniformly dampened because variations in moisture can lead to variation in color. The best way is to thoroughly soak the base the night before placement and then skip dampening prior to beginning work. You can also place a plastic vapor barrier on the subgrade before placing the concrete.

Two-Course Placement. The two-course method uses a base (or bottom) course of conventional concrete and a thinner top course of colored sand mix. Sand mix is a mixture of

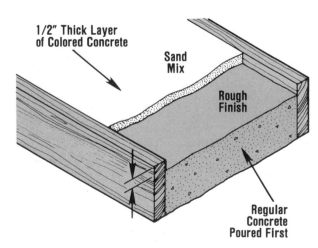

Details of two-course colored concrete construction

cement and fine sand without coarse aggregate. A mix of three parts fine sand to one part cement will produce a strong, durable top coat.

The colored top coat should be no less than 1/2" thick and not greater than 2" thick. If you plan to stamp or groove the colored top layer, be sure it is thick enough to accept the planned treatment. The sand mix topping offers another advantage in that it is extremely easy to work and finish and accepts stamping cleanly.

The base or bottom course is placed like an ordinary slab but stopped short of the form's top edge. You might also want to taper the base course near the sides to ensure all but a small area of exposed concrete is colored.

The surface of the base coat is not finished, but left rough to provide a better bonding surface for the top coat. This top coat can be placed as soon as the base course is stiff enough to support your weight and all bleed water has disappeared. Don't alternate between base and top courses. Place the entire base course first, thoroughly clean your equipment, then switch to top coat preparation. On bigger jobs this might mean a time lapse of several hours or even days. If this is the case, cure the base course as if it were a finished project, and treat the top coat application as a separate project. In this regard the work becomes strikingly similar to the resurfacing project covered on pages 160 to 163. The use of a brushable concrete bonding adhesive is recommended.

The top coat is floated and troweled in the standard manner.

Dry-Shaking

In this method of coloring concrete, special coloring compounds are dusted onto the surface of fresh concrete and worked down into the surface during the finishing process. These compounds consist of a mix of pigment, portland cement, and fine silica sands.

Always follow the manufacturer's precise instruction with regard to timing application. Normally, the dry-shake material is applied in two dustings. The first dusting takes place after

screeding, tamping, edging, jointing, and first floating. First floating brings some bleed water to the surface, and this water mixes with dry-shake material. Two-thirds of the recommended application amount is generally applied at this time. Even application is a must for consistent color tones.

Once the dry-shake material absorbs this surface moisture, it should be thoroughly floated into the top surface of the concrete. The remaining one-third of the dry-shake material is then applied at right angles to the first application, and the surface is refloated. Jointed and edged areas should be retooled after each floating. Troweling may be done after second floating if a denser finish is desired.

Painting

Most paints used on concrete and masonry surfaces are latex based, although epoxy resin and oil-based paints are also available. Alkali resistant additives are often formulated into latex paints intended for concrete and masonry to make them more durable. Antiskid paints containing fine sand or other abrasives are also available for steps and other traffic areas.

Time taken to read labels in the paint department of your local hardware store or home center is time well spent. Paint technology is constantly advancing, so buy a quality product and follow the application instructions precisely.

Most concrete and masonry paints and coatings are formulated for both above and below grade applications. Above grade surfaces must resist wind, rain, and weathering. Below grade surfaces are subject to groundwater exposure and naturally occurring salts in the soil.

Many specialty concrete paints and coatings are designed to prevent water penetration. Some products are breathable; they allow water vapor to pass, but stop liquid water. Breathable coatings are applied to exterior surfaces. They allow water vapor to travel through walls and floors and to prevent blistering due to vapor pressure or spalling due to the freezing of entrapped moisture.

The easiest way to apply paint to concrete or masonry is with a long nap (3/4" to 1") roller. Do not spread the paint too thinly, particularly on the first coat. The rough texture of concrete or block will ruin any brush, so use an inexpensive trim brush for touchup work. A whitewash brush is also a good tool for covering large areas quickly. Work when the surface is cool and shaded to prevent rapid drying of the paint.

Other concrete paints and coatings are designed to be impervious to both vapor and liquid water. These products are commonly used on interior surfaces. They prevent water vapor inside the building from entering the concrete or masonry materials.

Surface Preparation. Good surface preparation is essential for success. Because moisture within the concrete can cause some paints to blister and peel, fresh concrete must be cured for the period specified by the paint manufacturer. This can range anywhere from twenty-eight days to six months, depending on the product.

Never use curing compounds on concrete that is to be painted. They trap moisture within the concrete and also form a film that interferes with the bond between the paint and concrete.

Existing concrete must be clean and free of grease, oil, and efflorescence. See chapter 10 for details on cleaning concrete prior to painting. Additional information on portland cement-based coatings for concrete and masonry can be found in chapter 9.

Staining

Concrete stains are an excellent way to protect and beautify all types of concrete and masonry surfaces including poured and precast concrete, concrete block, common and faced brick, stucco, and natural stone. Stains blended with silicone acrylics offer deep penetrating coverage with outstanding durability and weather resistance. Stains also preserve the original surface texture of the concrete or masonry.

The concrete or mortar should be at least sixty days old prior to staining. The surface must be clean and completely free of all grease, oil, dirt, and release agents. Smooth surfaces should be etched using a solution of one part muriatic acid to two parts water (based on 20 baume acid strength). Apply the acid using a plastic sprinkling can and allow the reaction to continue for about ten minutes before rinsing with plenty of water. When properly etched, concrete will have a sandpaper feel when rubbed. Allow the concrete to dry for twenty-four hours before applying the stain. See chapter 10 for precautions when using acids or harsh cleaning chemicals.

Stains can be applied using a brush, roller, or airless sprayer. Apply liberally to allow for full penetration. Tightly mask any area not to be stained and work carefully along the masked line to prevent the stain from flowing under the masking tape or template.

EXPOSED AGGREGATE FINISHES

Exposed aggregate is perhaps the most popular of all decorator finishes. It offers a virtually unlimited selection of colors, textures, and aggregate size. Exposed aggregate finishes are attractive, rugged, slip resistant, and highly immune to wear and weathering. They are ideal for sidewalks, driveways, patios, pool decks, and almost all other types of slab work.

Materials for exposed finishes include naturally colored gravels and crushed aggregates such as quartz, marble, granite, and limestone. An aggregate size of 3/4" to 1" diameter is recommended for most work. Aggregate should be generally round or oval. Avoid using long, slivery materials.

Depth of exposure should be no more than one-third the diameter of the largest particles and one-half the diameter of the smallest aggregate in the topping material. There are three methods of obtaining an exposed aggregate finish: seeding, conventional, and top coating. The first two methods are well within the capabilities of most amateurs, but top coating should be left to experienced installers.

Seeding Method

In this method, the special colored or crystalline aggregates are seeded or pressed into the top of the fresh concrete slab. After sufficient setup time, the embedded aggregate is exposed by flushing and scrubbing off the top layer of concrete.

Seeding is not difficult, but it is time consuming. This is one job you should break into smaller sections. You might also want to apply a surface retarder to the slab to slow the curing process. This will give you more time to work in the aggregate and perform the flushing operation, but retarders are not essential for success.

Formwork and basic placement procedures are the same for all slabs. Concrete that is to receive a seeded exposed aggregate finish should be a little stiffer than normal. A 3″ maximum slump is about right. Place, screed, and bull float or darby in the usual manner, except set the finished height of the slab about 3/8″ to 1/2″ below the top of the forms to accommodate the height of the exposed aggregate.

Spread the aggregate uniformly over the surface by hand or use a shovel so that the entire surface is covered with a layer of stone. Embed the aggregate initially by tapping with a wood float, screed board, or darby. For final embed-

Pressing aggregate below the surface of the fresh concrete

ding, use a bull float or hand float until the appearance of the surface is similar to that of a normal slab after floating.

Exposure. Timing of the start of the aggregate exposure operation is critical to success. In general, wait until the slab can bear the weight of a person on kneel boards without indentation. Then brush the slab lightly with a stiff nylon bristle broom to remove excess mortar.

The next step involves fine pressure spraying and brooming. Special spray brooms can be rented for this job, but a two-person team using a garden hose and heavy push broom can do just as well. If brooming dislodges aggregate, delay

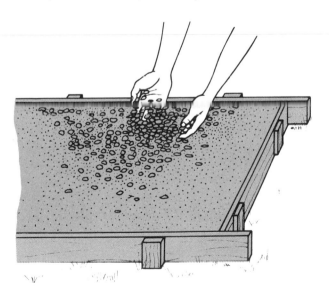

Seeding decorative aggregate onto the top surface of fresh concrete

Scrubbing away the top layer of cement paste to expose the decorative aggregate

Rinsing away excess cement paste from the aggregate surface

the operation until setup is more complete. Once brooming begins, continue washing and brushing until flush water runs clear and there is no noticeable cement film left on the aggregate.

Conventional Method

In this method, decorative coarse aggregate is used throughout the entire slab. The concrete is placed in the conventional manner and the top surface broomed and flushed at the appropriate time during setup. Although easier than seeding, the obvious drawback is the waste of expensive decorative aggregate.

When using this method, about 70% of all aggregates in the mix should be decorative coarse aggregate; that is, uniform in size, bright in color, closely packed, and evenly distributed throughout the concrete. Use a somewhat stiffer mix with a slump of 1" to 3". This will prevent the coarse aggregate from settling to the bottom of the slab. Follow the normal procedures for placing, screeding, and floating. Don't tamp or overfloat the surface because this will push the decorative stone too far beneath the surface.

Aggregate is ready for exposure when the water sheen disappears and the surface can bear your weight without indentation. If the scrubbing and flushing procedure dislodges aggregate or washes away too much cement paste and sand, delay the operation for a short period.

Top Coat Aggregates

A third method of creating an exposed aggregate finish involves placing a thin top coat of special aggregates over a base slab of plain concrete. This technique is commonly called *terrazzo construction*. The top coat can be as thin as 1/2" and contain decorative aggregates such as marble, quartz, or granite chips. With terrazzo top coats, the aggregate is exposed by a grinding, flushing, and scrubbing process. To prevent random cracking in the thin layer, brass or plastic divider strips are placed in a bed of mortar to act as control joints. These dividers also allow the use of different colored terrazzo mixtures in a variety of patterns. (While only a qualified terrazzo contractor should install the critical top coat, you can probably handle the work through to base slab construction.)

An innovative alternative to standard terrazzo topping uses synthetic aggregate chips made from recycled thermoplastics. The plastic aggregate is blended with portland cement to produce a finished topping coat that closely resembles terrazzo, with a polished surface and imbedded stone-like "chips" available in a variety of muted or bright colors. It offers outstanding resistance to wear, impact, and strong cleaners. The topping material is available as a premix containing both aggregate and concrete. Some ready-mix companies may also offer custom batching with plastic aggregate.

Cleaning and Sealing Aggregate Finishes

Even the lightest film of concrete residue can dull an exposed aggregate finish. After the slab has properly cured, it may require further cleaning. Commercial detergents and hot water will often do the job nicely, but some light abrasive scrubbing may be needed.

A dilute acid wash of one part muriatic acid to nine parts water can also be used to clean and brighten the aggregate. (See chapter 10 for precautions.) Delay the acid cleaning for at least two weeks after the job is completed. Acids may harm some limestone and marble aggregates, so run a test on some loose stone, watching for excessive foaming or discoloration. Wet the surface with fresh water before applying the wash, and flush all traces of acid away once it has dissolved the film.

If you wish, you can apply a clean protective coating to the cleaned aggregate to help maintain its bright appearance. Select a quality product with a base of acrylic resins. It will dry clear and stay clear.

Setting Small and Large Aggregates

Small size (3/4″ or less) aggregates, such as pebbles, colored gravel, and seashell fragments can be hammered into partially set concrete. Use a small board to evenly distribute the force of each blow as you set the aggregate. As with larger exposed aggregate the work begins after bull floating or darbying. Light scrubbing and flushing arc normally needed to clean and brighten the aggregate. Never use an acid wash on a seashell aggregate finish. Apply a clear sealer for lasting protection.

Larger cobblestone and river stone can also be set in the surface of slabs and step treads (see photograph on page 101). Although often considered an exposed aggregate finish, the process more closely resembles setting pavers in a mortar bed (see chapter 14). In this case, the mortar bed is the freshly placed slab that should not be allowed to set up too stiffly before setting the stone. Hammer or press the stone well into the surface so more than 50% of the stone is embedded. Cover the stone with a protective board to prevent chipping or cracking while hammering. Brush the area between the stone with a small hand broom to create clean, even joints.

TEXTURED FINISHES

Many interesting and functional textures can be created using floats, trowels, or brooms or by applying splatter coats, rock salts, or sparkling abrasive granules.

Swirls

Swirled finishes create visual interest and surer footing. They are made using hand floats or trowels once the concrete is screeded, bull-floated, or darbied, and has lost its shine. Wood floats, magnesium floats, and steel trowels produce coarse, medium, and fine textures, respectively. The hand float or trowel is worked flat on the surface using pressure in a semicircular or fanlike motion. Patterns are worked up using a

Hammering small-size decorative aggregate into partially set concrete

A swirled finish created by using a hand float or trowel

series of uniform arcs or twists. The slightly raised ridges of the swirled pattern can be damaged during the early stages of curing. Use a light mist as opposed to a pressurized spray when wetting down the slab, and never drag burlap or plastic sheeting across the surface of the work.

Brooming

Pulling a dampened broom across the surface of freshly floated or troweled concrete is another way to create an attractive, nonslip surface. Coarser textures are produced using stiff-bristled brooms on newly floated surfaces. They are excellent for steep slopes or heavy traffic areas.

Medium to fine textures are created with soft-bristled brooms on floated or troweled surfaces. Sidewalks and driveways should be broomed at right angles to the direction of traffic. Rinse the bristles often to keep the tips clean.

And don't limit yourself to straight line patterns. Wavy and curved patterns offer interesting alternatives.

Travertine Finishes

Travertine finishes are created by applying a dash coat of cement paste and fine sand over freshly finished concrete. Use one part cement to one part fine sand, and add water until the mix has the consistency of thick paint. Yellow or other pastel pigments can be added to the dash coat. Apply the mortar by heavily loading the end of the brush and flicking it onto the surface. The result should be a splotchy finish with plenty of ridges and depressions. Allow the dash coat to set up slightly before flattening with a trowel. The final finish will have smooth high spots and coarsely grained depressions.

A travertine texture

A straight, coarse bristle-broomed finish provides excellent traction.

A wavy broomed texture

Rock salt textured surface

Walkways, driveways, and patios can be scored with a simulated flagstone pattern.

To produce a travertine finish, apply a dash coat of mortar, allow it to partially set, and flatten raised areas with a trowel.

A similar effect can be created by pressing or rolling rock salt into the surface of freshly floated or troweled concrete. Leave the tops of the salt grains exposed. After the concrete is cured, dissolve and dislodge the rock salt by washing and light scrubbing. The result is a deliberately pitted surface.

Travertine or pitted finishes are not recommended for cold climates where water trapped in the depressions would expand upon freezing and crack the finish.

GEOMETRIC PATTERNS

Designs stamped or scored into the fresh concrete surface of walkways, driveways, and patios can be used to simulate random flagstone or ashlar patterns.

One method involves embedding 1″ wide strips of 15-pound roofing felt in the concrete. After initial floating or darbying, the felt strips are laid on the surface, patted flush, and then floated over. If desired, dry-shake pigment can be applied at this time, or the concrete can be left its natural color. Once the slab stiffens sufficiently, the felt strips are carefully removed to expose the simulated mortar joints.

The same effect can be created by pressing or scoring grooves into the surface with a piece

Use a Jointing Tool to Groove the Concrete to Look Like Flagstone.

A jointing tool or piece of bent pipe can be used to score the surface of partially set concrete.

of 1/2″ or 3/4″ diameter copper pipe. The pipe should be about 1-1/2′ long and bent into a flat "S"-shape for easier handling. Scoring must be done when the concrete is still plastic—just after first floating or darbying. Pick out any coarse aggregate that cannot be pushed beneath the surface. Once the pattern is complete, float and trowel to the desired finish. Touch-up the grooves as needed during finishing.

Stamped Patterns

Special stamping pads available at most large rental centers can be used to imprint simulated paving brick, stone, tile, or other patterns

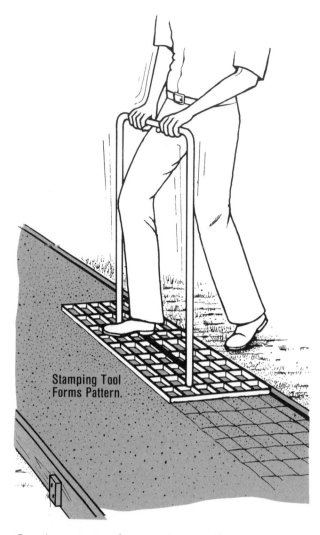

Creating a stamped pattern in partially set concrete

into partially set concrete. When used along with coloring pigments, the final surface effect is very authenic looking and durable.

Concrete that will receive a stamped finish pattern should contain small coarse aggregate such as pea gravel. Finishing follows the usual procedures; however, do not trowel the surface more than once.

After the surface is floated and troweled to the desired texture, carefully position the stamping pad or pads on the concrete surface. Some patterns require two pads placed side by side to correctly produce and align the desired pattern. When a large area must be covered, such as a driveway, up to eight separate stamping pads are positioned at one time.

Simply step on the pad to stamp the design to a depth of about 1″. A hand tamper can also be used to ensure proper depth penetration. A tool similar to a brick mason's jointer is then used to dress the edges and create some artificial imperfections. A small hand stamp might be needed to complete the pattern at the slab edges or in tight areas too small for the larger pad. Timing is critical because all stamping and jointing must be completed before the concrete sets too hard.

Permanent Dividers

Permanent divider strips made of rot-resistant or treated lumber, plastic, metal, or masonry pavers can also be used to create geometric patterns like diamonds, triangles, and squares. These materials can also serve as borders.

Incorporating permanent dividers into slab work has several advantages. It breaks the work down into smaller more manageable sections. The dividers also act as permanent control joints to minimize cracking. Each section can be finished in the same manner, or alternate surface treatments can be used.

Redwood, cedar, or cypress are all naturally rot-resistant woods. Plastic strips may be too flexible to install straight unless they are nailed or stapled to the top of wood strips securely staked to the subgrade.

Concrete masonry, brick, or stone divider strips and borders may be set in a sand bed with or without mortared joints. For more permanent

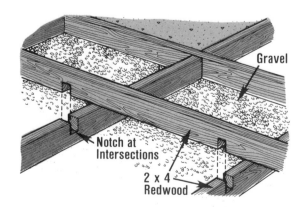

Details for a gridwork of permanent wood dividers in a patio or large walkway

This walkway combines an exposed aggregate finish with permanent wood strips and stretches of straight concrete.

work, masonry units should be set in a mortar bed, with all joints mortared. Refer to chapters 14 and 17 for more information on paving with brick, concrete pavers, and flagstone.

ABRASIVE GRAIN FINISHES

Special abrasive grains, such as silicon carbide and aluminum oxide, can be seeded into the surface of fresh concrete to create a long-lasting, nonslip finish. The application method is similar to dry-shake coloring, as described earlier. The grains are spread uniformly over the surface and lightly troweled into the very top layer of fresh concrete.

Silicon carbide grains are black and create a "sparkling" finish that is particularly attractive under artificial lighting, as on a patio or porch deck. Aluminum oxide powder can be gray, brown, or white. It does not create a sparkling finish.

Chapter 7

CONCRETE FOUNDATIONS, WALLS, AND STEPS

All types of concrete and masonry walls share one design characteristic: They all must be built on a strong, level concrete footing. This concrete footing, or footer as it is sometimes called, is the foundation of the wall. The footing provides a stable base for the wall, which is centered on the footing and tied to it by the use of a keyed form (concrete walls) or mortar bed (brick, block, and stone walls).

Placing a footing for a large foundation. Note the use of a long extension chute. Most jobs tackled by home-owners are not this large.

FOOTER DESIGN

A typical footer is twice the width of the wall it supports and equal in depth to the wall's width, or 8", whichever is greater. For example, a footer for a foundation wall constructed of 10"-wide concrete block should be 20" wide and 10" deep. The footer should rest on a 6" gravel base set below the frost line.

Frost Line Considerations

The frost line is the depth to which the ground normally freezes solid during the winter months. By setting the footing at a depth below the frost line, you eliminate the movement caused by freeze–thaw cycles. The frost line depth varies greatly throughout the United States. It can range from 0" to 10" in the south and west to upwards of 80" in the extreme northern sections.

Frost line depth is always a major factor when setting footings for foundation walls, pier foundations, retaining walls, tall screen walls, and any concrete or masonry structure that bears significant weight.

However, digging a 3'-deep trench to place a footer for a low garden wall or tree well project is not practical or necessary for success. Because local climates and conditions vary greatly, general recommendations for dealing with frost line

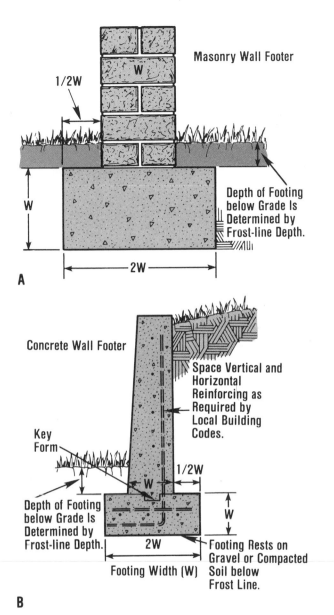

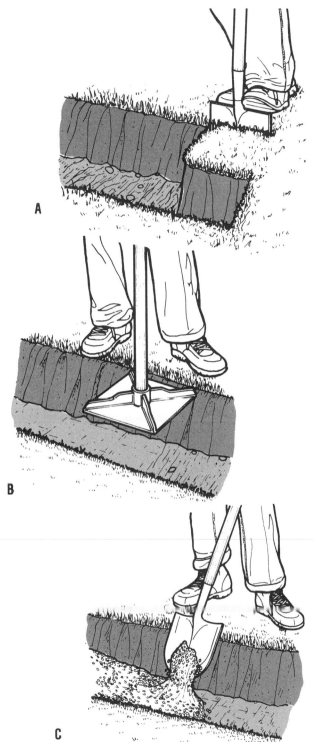

A

Masonry Wall Footer

1/2W

W

W

2W

Depth of Footing below Grade Is Determined by Frost-line Depth.

A

Concrete Wall Footer

Space Vertical and Horizontal Reinforcing as Required by Local Building Codes.

Key Form

1/2W

W

Depth of Footing below Grade Is Determined by Frost-line Depth.

2W

Footing Rests on Gravel or Compacted Soil below Frost Line.

Footing Width (W)

B

Typical footer designs used with (A) masonry walls and (B) concrete walls

depths are impossible. Always consult with local building authorities to determine the frost line depth in your area. Explain your project plans and follow their recommendations to the letter. Local building departments might also have design specifications and code requirements that differ from those suggested in this book. It is your responsibility to know and follow all local codes.

B

C

Steps involved in constructing an earth footer form: (A) digging with a square face shovel, (B) tamping the base, and (C) adding a gravel or crushed stone base (when needed)

Constructing Footer Forms for Walls

Lay out the footer location using batter boards as described in chapter 4. The layout must be extremely accurate. The footer must run straight and true if the wall is to run likewise. You will not be able to correct mistakes during the construction of the wall. Pay particular attention at corners. An out-of-square foundation will result in an out-of-square building.

Earth Forms. If the soil is firm enough to hold its shape when filled with wet concrete, consider using an earth form.

Carefully follow the layout lines, digging to the correct depth and width using a square-faced shovel. Remember that increasing width and depth beyond what is called for will waste significant amounts of concrete. Firmly compact the soil using a tamper and add the 6″ gravel or crushed stone base to ensure good drainage.

Screed guides, leveled and staked at the correct height, will help in striking off concrete in earth forms. After the screeding pass, these guides are removed and the gaps filled with extra concrete and leveled with a float or trowel.

Wooden Forms. In loose soils, construct strong wooden forms using sturdy 2″-thick lumber, stakes, and nails. Form boards must be parallel and level. Fill in any gaps at the base of the forms with stone—never use soil for footers. On slopes, construct a stepped footing to minimize the amount of concrete used. Stepped forms should rise no more than 2′ per step and overlap

Stepped formwork for footers

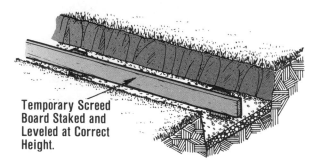

Temporary Screed Board Staked and Leveled at Correct Height.

Position of a temporary screed guide for earth-formed footers

A complete stepped footer

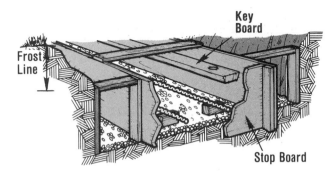

Formwork for keyed footer construction. Note the hole in the key board for the insertion of vertical rebar.

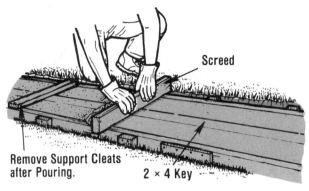

Placing concrete in footer forms and screeding to the top of the forms and key board

at least 2' at each step. The same design principle can be used in earth forms as well.

Keys and Reinforcement. It is recommended that steel rebar be positioned one-third of the way up from the base of the concrete footer. If the footer is being used as the base for a poured concrete wall, a keyed slot should be designed into the top of the footer. This keyed slot should be about half the width of the finished wall. It is formed by positioning a key board down the center of the form as shown. The top of this key board is level with the top of the formwork. The board is removed after the concrete is poured and set up. With earth forms, the screed guide might interfere with the positioning of the key board. If this will occur, the key board is positioned after the screed guide is removed and the concrete is still soft enough to accept the key board.

If vertical rebar reinforcement is going to be used to help tie the footer and wall together and strengthen the wall, drill holes in the key board to make positioning it during the pour easier.

Placing the Footing

Compared to slab work, placing a footing is simple, provided you don't have to haul the concrete a great distance. Double check all formwork, reinforcement, and so forth. If the footer abuts an existing structure, install an isolation joint at this location. Dampen the subgrade just prior to beginning placement.

A slightly wetter mix will help the concrete flow a little easier. Work from one end of the form to the other. Fill all voids and work out the air pockets by working the end of a rod in and out of the concrete. You can also rap lightly on the side of the forms to help settle the concrete in

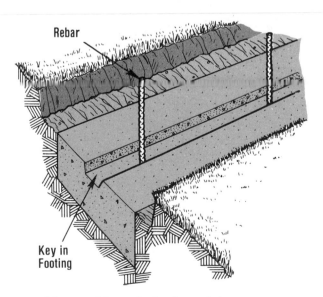

Keyed footer with vertical rebar reinforcement

deeper forms, but don't overdo it because the large aggregate might settle out of the mix.

Screed the concrete level and remove any screeding guides. If needed, install the key board when using earth forms. With wooden forms, work around the key board that is already in position. Install any vertical rebar through holes in the key board, and temporarily brace it as needed.

The top surface of the footing requires little more than screeding and a light hand floating. A slightly rough finish is actually preferred. Remove the key board as soon as the concrete has set up to hold its shape.

Cure the footer for at least three to four days before removing the forms and continuing work.

CONCRETE WALLS

The clean, smooth lines of a poured concrete wall adapt to any landscape or setting. Easily formed in curves, straight lines, or irregular shapes, walls of concrete are strong and durable with most applications requiring no more than 8" widths.

Concrete walls are widely used as foundations for many types of structures. Concrete walls can also be used in garden or patio areas as decorative screens, borders, planter walls, tree wells, or retaining walls to control erosion or landscape an area.

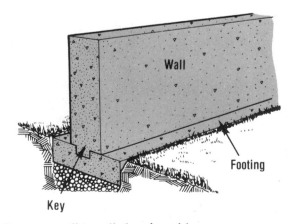

Concrete wall installed on keyed footer

Positioning poured concrete wall formwork on the footer. These are prefabricated wall forms for commercial work.

Building the Forms

The most important step in constructing a strong, attractive wall is building strong, accurate forms. Wall forms must be strong enough to withstand the great pressure exerted by the wet concrete; any failure in the forms will be disastrous. Keep in mind that building and aligning the forms for a poured concrete wall usually takes much longer than pouring and finishing the concrete.

A straight wall form is constructed of 1/2", 5/8", or 3/4" plywood sheathing, studs, spacers, ties, and (for larger, heavier walls) wales. Sheathing forms the mold, while studs back up and support the sheathing. Spacers set and maintain spacing and support the form prior to the pour. Wire ties snug the form and resist the pressure of the wet concrete. Wales align the form and brace the studs in forms more than 4' to 5' high. Two horizontal wales are sufficient for most forms, but they should not be spaced for greater than 30" on center.

For lower, lighter walls, it is possible to cast the wall at the same time you cast the footer. When using this method, stake the form firmly to the ground. The fresh concrete will tend to lift the entire form out of position. This trench formwork combination is also used for certain retaining wall designs. Larger walls always re-

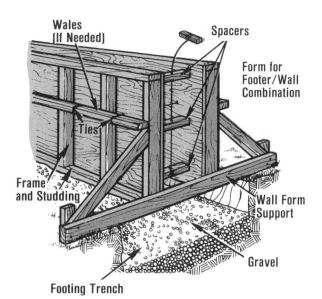

Formwork for placing smaller concrete wall and footer in one pour. This same setup can be used for slab on footer retaining wall construction.

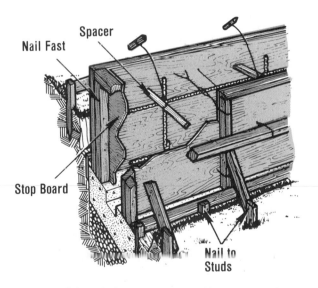

Formwork for placing concrete wall on existing footer. Once again the design principles can be used for freestanding or retaining wall construction.

quire separate pours for the footer and wall, with the wall keyed to the footer as shown in the illustration on the opposite page.

Form Strength and Pour Rate. Fresh concrete acts like a liquid, which means it develops a hydrostatic pressure that is directly related to its height. In slab work its downward pressure is minimal since the concrete is only 4″ to 6″ above the ground. But when constructing a wall, fresh concrete may be several feet off the ground.

With wet concrete, the hydrostatic pressure pushing outward against the lower sides of the form will be equal to the total weight of the concrete, plus the forces generated by dropping the concrete a certain distance into the form. For an 8′ wall, this pressure can exceed 1200 pounds per square foot. To withstand such pressures, forms must be extremely strong and stable. Fortunately, controlling the rate of pour will significantly reduce hydrostatic pressure.

If the concrete is placed slowly in layers and given time to set up and stiffen, it will be able to carry its own weight. The same 8′ wall placed at a rate of 1 vertical foot per hour will generate only 150 pounds per square foot, a pressure easily contained by soundly constructed wall forms. Table 7–1 lists plywood thickness and stud spacing for various vertical pour rates.

If you plan to use ready-mix delivery, discuss pouring rates with the ready-mix plant before you begin your form construction. Having a mixer on site for several hours incurs additional expense. You'll want to reach the best possible balance between the costs of the forms and the cost of the ready-mixer's time. Be sure the driver is aware of the agreed upon pour rate and the time expected to spend on site. And once the pour begins, do not increase the rate for any

TABLE 7–1: POUR RATE FOR VARIOUS SHEATHING AND STUD SPACINGS

Sheathing Thickness	Pour Rate	Stud (2×4s) Spacing
1/2″	1 ft/hr	13″
3/4″	1 ft/hr	18″
1″	1 ft/hr	22″
1/2″	2 ft/hr	11″
3/4″	2 ft/hr	16″
1″	2 ft/hr	20″
1/2″	4 ft/hr	9″
3/4″	4 ft/hr	12″
1″	4 ft/hr	16″

reason. A collapsed wall form during a pour is the ultimate concrete disaster.

To construct the forms:

1. Build the form in sections, using 2×4s laid on edge to construct frames that measure the height of the wall and no more than 8' in length. Nail 2×4 studs into each frame, spacing them on center according to the distances specified in table 7–1.
2. Nail plywood sheathing to the frames. If needed, mark off the position of the wales and toenail them to the studs.
3. To install wire ties, drill 1/8" holes on either side of the wales or studs. Tilt two sections upright, face to face, spacing them at the desired wall thickness. Run a piece of wire through opposite holes in the form and around wales or studs. Twist the ends together to form a loop. Insert a properly sized spacer near the tie, and tighten down the tie by using a stick to twist the tie snug as shown. Remember to attach pull wires and around spacers so they can be removed as the pour is made.
4. Add on sections by nailing frames together through adjacent studding. The running length of the form should be slightly longer than the finished wall so

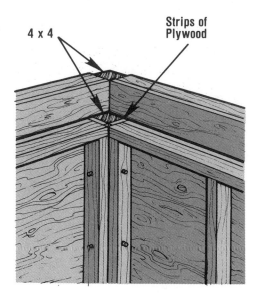

Constructing corner formwork for concrete walls

that a stop board can be installed as shown.
5. Center the completed form over the footer, making certain it is plumb. Stake, brace, and nail the form firmly in place.

Corners and Openings. Form sections are joined at the corners by nailing end studs to 4×4 posts as shown. Strips of plywood are nailed to the inside 4×4 post to locate it flush to the sheathing.

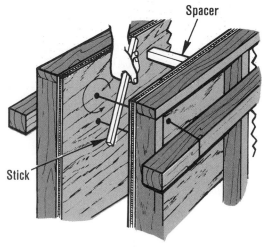

Installing and tightening wire ties

Placing wooden window "buck" in position. When the outer wall of the formwork is positioned, the buck will be sandwiched between the form walls.

A wooden form, called a *buck*, can be nailed between the two formwork walls to produce an opening for a window or door. When the concrete is placed, it surrounds the buck, which is removed after all other forms are removed. Apply clean oil or a release agent to the outer surface of the buck. When an opening extends to the top of the concrete wall, there is no need for overhead support within the wall. The same is still true if the opening is within the wall but is less than 2′ wide and runs parallel to the floor joints. However, when the opening is wider than 2′ or is subject to heavy loads, additional reinforcement should be provided above the wall. The easiest method is to suspend 3/8″ rebar rods in the form 2″ to 3″ above the top of the buck.

Curved Formwork.

Gently curving forms can be created using strips of kerfed plywood. Curved formwork reduces the amount of horizontal support that can be used, such as wales and framing lumber. Never eliminate the horizontal framing lumber at the base of the wall forms. Kerf this piece as needed to bend it. Nail the sheathing to the base framing and firmly stake the base framing to the ground as shown. Increase the number of vertical framing members used, and solidly brace and stake each to the ground.

More severe curves can be formed using unkerfed 1/8″ or 1/4″ plywood bent and secured to a series of curved horizontal ribs. The ribs are cut from 3/4″ plywood using a saber saw. They should be a minimum of 4″ wide, and spaced every foot up the wall as shown. This method eliminates the need for vertical supports in the curved section.

To help prevent the plywood from deforming or cracking the ribs, clamp it to 2×2 wood blocks positioned at the midpoint of the curve as shown. These temporary blocks reduce the stress placed on the ribs and plywood until the plywood can be securely nailed to the ribs. Begin by nailing at the ends and working in stages to the midpoint of the curve. Remove the blocks and clamps prior to nailing at the midpoint. Nail the curved form into the vertical supports of the adjoining straight wall forms. Brace the curved section as needed.

Cap Forms.

An overhanging cap can be created by nailing furring strips to the outside of

Kerfed strips of plywood are used to create curved forms for low walls. Strips butt tightly on top of one another and are nailed to vertical supports.

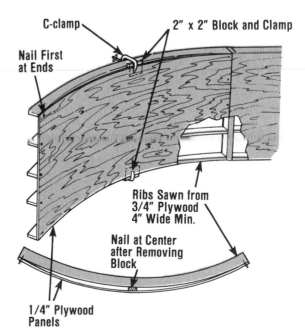

Details for curved formwork using thinner unkerfed plywood and solid horizontal support ribbing

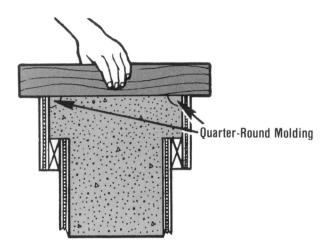

Details for forming an overhanging wall cap

Pouring a large concrete wall. A splashboard is used to direct the concrete. Note the use of large clamps to hold a window buck in position.

the forms to widen them slightly. The wall portion of the form is filled with standard concrete mix, but the cap portion of the form should be made with a mix containing more sand and no large aggregate. This will allow for a smooth, dense finish. Decorative molding positioned inside the cap form creates more elaborate edging effects.

Pouring the Wall

With the form properly mounted on the cured concrete footer, tie the wall rebar into the existing footer rebar. Coat the insides of the forms with clean oil or release agent. This is particularly important in wall work.

The concrete mix should use a moderate amount of water. It should not be so loose that it finds its own level. When this is the case, aggregates settle at the base of the wall and overall strength is weakened. Extremely dry mixes require too much labor to push and pull the concrete in place, and smooth surface finishes are very difficult to obtain.

If you are using ready-mix, consult the plant about the appropriate mix proportions for your job. If you are machine mixing yourself, consider using a 1:2-3/4:4 mix for foundation and service walls not exposed to the weather. A 1:2-3/4:3 mix ratio should be used when the wall is exposed to the weather or when a smooth surface finish is desired.

Pour the wall in horizontal layers, working at your predetermined pour rate. Exceeding this rate is rarely a problem when mixing your own concrete, but it can be when using ready-mix. Once you begin filling the form, stopping for any appreciable length of time is out of the question. The uppermost layer of concrete must always be fresh so the succeeding layer can be worked into it smoothly. This is the only way to obtain a smooth, continuous wall. If layers are allowed to set up too firmly, cold (nonbonded) joints will result.

Depending on the amount of concrete to be poured and the people available to do the job, it might be necessary to construct the wall in sections by using a movable stop board. Drill holes through the stop board so that it can be moved along the wall without cutting the rebar.

Begin at the ends of the form and work toward the center. Avoid excessive handling of the concrete. Reposition the ready-mix chute as needed or, if possible, set up a ramp to wheel the concrete into position. Set up a splashboard to help direct the pour and control spillage.

Remember to remove the wooden spacers between the form walls as the pour is made. Work the concrete against the sides of the form and around the reinforcement as each layer is poured. Strike the sides of the form with a

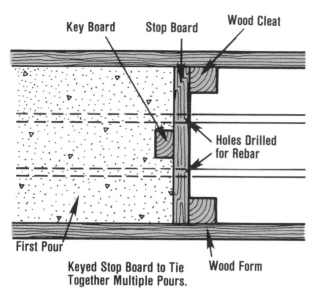

Details for installing a stop board inside concrete wall formwork

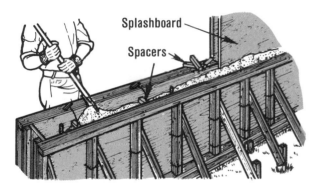

Placing concrete in layers for low wall construction

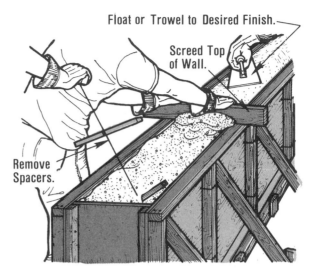

Screed and finish the top of the wall. Remove temporary spacers.

Note: When the wall is serving as a foundation for a wooden structure, anchor bolts should be carefully positioned in the concrete when it reaches the proper stiffness. Position the bolts down the center of the wall, spacing them at proper intervals for the intended construction. Precisely spacing and aligning anchor bolts at this time will prevent major headaches later in the construction of your building.

If the wall is placed in sections using a key board, allow it to set twelve to twenty-four hours before removing it and continuing to work. Ap-

hammer or mallet to settle the concrete against the sides and bring fine aggregate, sand, and cement to the surface.

Renting an internal concrete vibrator can help you achieve the best results. The shaft of the vibrating tool is placed deep within the form. The vibrating action consolidates the concrete and brings fines to the surface. With a vibrator, slightly stiffer mixes can be used without sacrificing smooth finishes.

Once the top of the form is reached, screed, float, and trowel to the desired finish. The finish on the face of the wall depends on the surface finish of the formwork. Of course, the cured wall can be given any number of surface treatments, such as stucco, or masonry coating finishes.

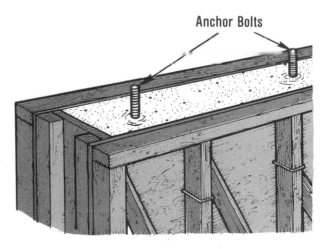

When needed, anchor bolts for wooden sills and so forth should be carefully positioned in the fresh concrete.

ply concrete bonding adhesive at the joint to provide a leakproof juncture.

Remove the wall forms after the concrete has cured for at least three to four days. Cut the wire ties flush to the surface of the concrete. Fill in any honeycombed areas with a mix of cement and sand. Cover the tips of the wire ties to prevent rusting and staining.

CONCRETE RETAINING WALLS

Retaining walls are used to prevent soil erosion of sharply sloping lawns and gardens. Retaining wall designs include straight slab on edge, straight slab on footer, and truncated pyramid designs. In all cases, strong formwork and provisions for drainage are the keys to success. This is another project that should always be cleared with the local building authorities during the planning stages. Many areas specify that retaining walls above a certain height (usually 4') must be designed by a licensed civil engineer.

Slab on Edge Retaining Walls

Slab on edge retaining walls are the best design when the wall must extend into the ground 2' or more to reach the frost line. Common dimensions for slab retaining walls are given in the illustration below. These walls will serve well under average conditions; that is, the soil is fairly firm and well drained and the slope is not too severe. If the hillside is high in clay content and subjected to heavy rains, stronger designs might be necessary.

Begin by cutting a level shelf into the hillside at the proposed wall location. Take the time to create adequate working room.

If the soil is firm, consider an earth form (trench) for the below-grade portion of the slab wall. You can then construct forms for the above-grade section of the wall, saving considerable time and materials. As shown in the illustration, forms can be similar to the plywood and framing designs discussed earlier. They can also be made using 2×12 or 2×10 planking set on edge for the form walls.

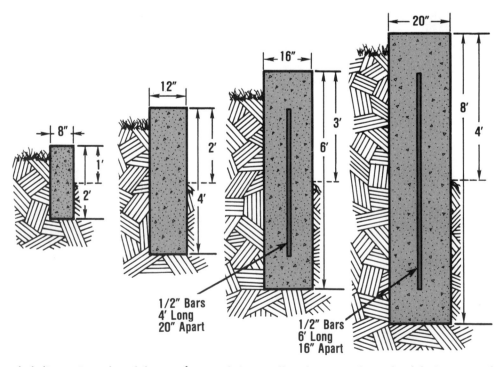

Recommended dimensions for slab on edge retaining walls. The same footerless design can also be used for freestanding walls

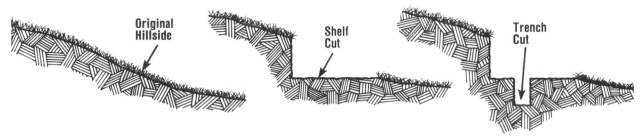

Steps in preparing the site of a hillside retaining wall.

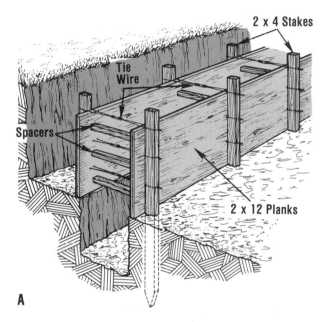

A

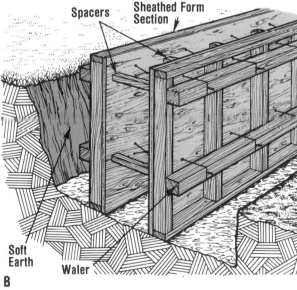

B

Formwork for slab on edge walls using (A) partial earth forms and (B) full form for walls in loose soil. Fill in and grading will be needed on both sides of the finished wall.

As with all walls, pour at a rate that will not overload the forms. A 1:2-3/4:3 mix ratio should be used. As discussed earlier, work in layers, placing in sections if needed. Screed level with the top of the forms and finish as desired.

Slab on Footer

Slab on footer retaining walls are most economical when there is no need to dig more than 6" to 12" to reach firm, frost-free soil. The slab or wall does not have to be centered on the footer as shown; many "L" and inverted "T" designs are possible. Maximum resistance to toppling is obtained when the footer is beneath the hillside. But this design requires maximum backfilling. Install reinforcement rebar as shown.

With this type of retaining wall, the footer is cast at the same time the wall is placed. This is exactly the same as the formwork design used for lighter, lower walls shown on page 87.

Fill the footing section first. Hold off placing the vertical wall for thirty minutes or so until the footer stiffens sufficiently to support the vertical load without deforming. When this point is reached, you can slowly place the vertical wall in layers as described earlier.

Truncated Pyramid Walls

Pyramidal retaining walls resist the push of the hillside by their sheer mass alone. You can therefore use the least expensive concrete mix possible 1:3:5, and larger, clean aggregate. Movement from freeze–thaw cycles is not a major concern unless the frost line depth exceeds 1' or more. Even then you can excavate to within 6" of this depth and proceed safely.

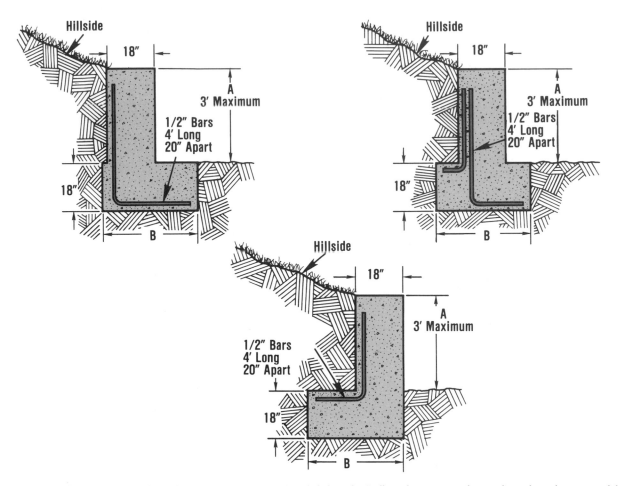

Dimensions for integral slab on footer retaining walls. Slab height "A" is always equal to or less than footer width "B". Slab and footer thickness are equal.

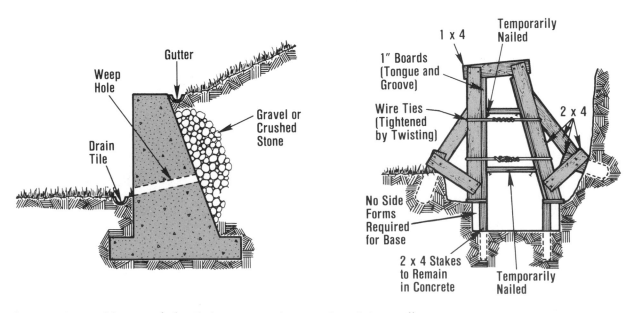

Construction and formwork details for truncated pyramid retaining walls

TABLE 7–2:   RETAINING WALL CONSTRUCTION DATA

Exposed Wall Height (A)	Top Thickness (B)	Distance from Ground to Base (C)	Distance from Top to Base (D)	Base Depth (E)	Base Width (F)	Outside Base Extension (G)	Inside Base Extension (H)
12″	6″	4″	16″	6″	14″	3″	3″
18″	6″	6″	24″	6″	18″	3″	3″
24″	7″	8″	32″	8″	24″	4″	4″
30″	7″	10″	40″	10″	28″	4″	4″
36″	8″	12″	48″	12″	36″	6″	6″
42″	8″	14″	56″	12″	40″	6″	6″
48″	9″	16″	64″	12″	44″	6″	6″

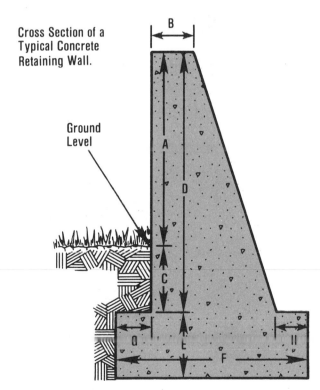

Cross Section of a Typical Concrete Retaining Wall.

Ground Level

Dimensions of typical truncated pyramid retaining walls. See table 7–2.

The dimensions of the retaining wall will vary according to the wall height, as shown in table 7–2. The design shown here does not require steel reinforcement since the width of the base and the weight of the wall provide adequate support.

Details of the retaining wall form are shown on the previous page. Once again, footer and wall are cast in one step. If the soil is sufficiently firm, use an earth form for the footer portion.

Follow the placement procedures outlined earlier. Because of the pressure created by the sloping aggregate, cure the concrete for at least seven days before removing the forms. Once the forms are removed, tamp crushed stone fill into the space behind the wall. Fill the top foot or so with topsoil, providing a gutter depression along the wall for better drainage.

Retaining Wall Drainage

All retaining walls higher than 2′ above ground must be drained to prevent water from collecting behind them. Water pressure and frost heave from behind the wall can quickly push the wall away from the hillside. Your autumn project might be well on its way to ruin by spring.

There are two ways to drain a retaining wall. The first is to provide weep holes through the wall. Weep holes are made by inserting short lengths of 2″-diameter plastic drain tiles in the formwork prior to placing the concrete. The pyramid retaining wall on the opposite page shows the use of weep holes. The plastic pipe is sloped downward from back to front. The pipe exits the wall face 2″ to 4″ above ground level. Gravel or

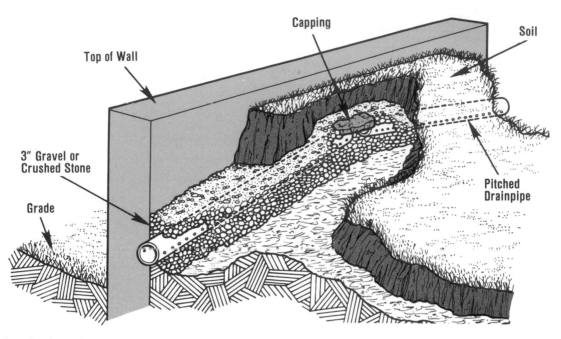

Providing for lateral drainage behind a retaining wall

crushed stone is used as backfill behind the wall to aid drainage, and drain tile gutters are placed at the base of the retaining wall.

The second type of drainage is lateral drainage. In this method, drainpipes are placed behind the wall during the backfilling process. As shown in the illustration, fill and tamp the area behind the wall with soil to a height of approximately 1-1/2' above grade at the center of the wall and 1' above grade at the wall ends. Add a 3" layer of crushed stone or gravel about 1' wide along the entire backside of the wall. Lay the lengths of plastic pipe in place. They should pitch downward from the center to the sides as shown. Use 3"-diameter pipe with holes drilled at regular intervals in the pipe's side. Extend the pipes beyond the ends of the wall. Cover the pipe with a 3" layer of crushed stone, and then backfill to the top of the wall.

CONCRETE FOUNDATIONS

The three most common concrete foundation designs are poured wall foundations, slab foundations, and pier or pillar foundations. The steps involved in placing a poured wall foundation have already been covered earlier in the chapter under concrete walls. Constructing a poured wall foundation is a major undertaking for even the smallest of structures. Fortunately, most foundations for garages, storage sheds, cabins, and other smaller structures are either slab or pier/pillar. Constructing these designs is well within the skill level of most do-it-yourselfers.

Slab Foundations

Slab foundations are simple and inexpensive. They eliminate the need for major excavation and foundation walls. Three variations of slab foundation construction are used: desert, wet/warm area, and cold area. As the illustrations show, all combine basic slab construction with a perimeter footing.

For any type of slab foundation, stake out the site with batter boards (see chapter 4), remove the top layer of sod and soil, and dig the footer trench as needed. The footer may be cast with earth or wooden forms depending on soil consistency and the trench depth required. The slab section is formed with wooden formwork.

Placement and finishing is the same for all basic slab work (see chapter 5). Control joints are not used. Normally, a smoother troweled finish is desired. Foundation anchoring bolts should be inserted headdown in the wet concrete along the edge of the slab. These bolts must be properly spaced and aligned according to the building plan. As emphasized earlier, this is not a haphazard operation but a critical step in the building's construction.

Desert Slab Construction. This type of foundation is constructed the same way as any slab, such as sidewalks or patios, except that a shallow footing is poured with it. This slab rests directly on the ground.

1. Remove the topsoil and dig the trench as shown.
2. Install and insulate any service pipes needed for water, sewer, and the like.
3. Construct the forms; stake and nail them firmly in position. Note how anchor bolts can be temporarily positioned for the placing of concrete.
4. Mix and place the concrete in the footer and slab forms in one continuous operation.
5. Screed, float, and trowel the concrete as if it were a sidewalk or patio.
6. When constructing thicker slabs, use 4" of gravel base to save on the amount of concrete used. For larger slabs use a length of 4×4 lumber as a temporary screed guide.
7. Cure the concrete for three to four days.

Wet/Warm Area Construction. This type of slab is used in wet/warm areas that are free of freezing problems. The footer and slab sections are poured in two steps, and a moisture barrier is installed.

1. Construct and install two separate forms as shown. Use short lengths of board to hold the inner form in position. Fill the inner form with an 8" gravel base and install a moisture barrier, such as 6-mil plastic sheeting, over the gravel.

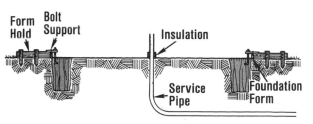

Formwork details for basic desert slab foundation construction

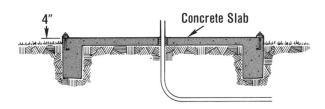

Cross section of finished desert slab construction

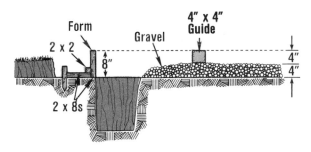

Formwork for thicker slab foundations using a 4" gravel or crushed stone base

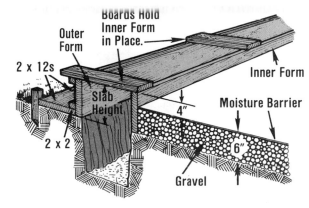

Formwork details for wet/warm weather slab foundations

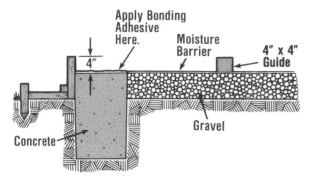

Preparing to pour the slab portion of a wet/warm weather slab foundation. Note the use of a 4×4 screed guide.

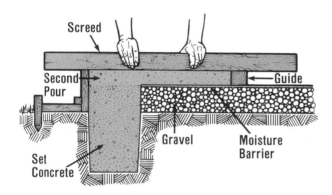

Placement and screeding details for a wet/warm weather slab foundation

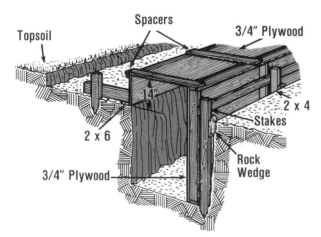

Footer details for cold weather slab foundation construction

2. Place the concrete in the footer trench 4" short of the top of the form. Allow the concrete time to set up and retain its shape (one to two days).
3. Remove the inner form. Apply a coat of concrete bonding adhesive to the exposed footer surface.
4. Pour, screed, and finish the balance of the slab form, removing the temporary screed guides as you proceed.
5. Cure the concrete for three to four days.

Cold Area Construction. The footer section of this slab must extend below the frost line, and the slab itself must be insulated with rigid board insulation.

1. Dig a 2'-wide footer trench to a depth several inches below the frost line. In firm soil, one side of the trench can act as a form. Construct and install the outer and inner form walls as shown.
2. Pour the concrete to the top of the form, vibrate or rod out air pockets, screed level, and install anchor bolts.
3. Once the footer has set up (one to two days), remove the footer form and install 4" rigid insulation board on the inner side of the concrete foundation. Use soil to hold the insulation in place.

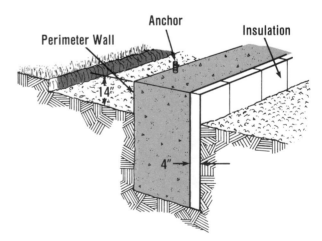

Position of 4"-thick rigid insulation on slab footer for cold weather areas

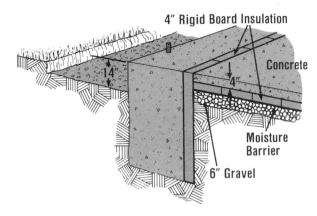

4" Rigid Board Insulation

Concrete

14"

4"

Moisture Barrier

6" Gravel

Subgrade, moisture barrier, and rigid insulation details for slab portion of a cold weather slab foundation

4. Lay down a 6" gravel base, cover with a moisture barrier, and lay rigid board insulation on top of the moisture barrier as shown.
5. Place a 4"-thick layer of concrete over the insulation board, screeding and finishing as before.
6. Cure the concrete for three to four days.

Pier Foundations

A pier or pillar foundation consists of individual concrete footers on which the piers or pillars are built. These piers or pillars can be made of concrete, concrete block, or brick. Anchor bolts or plates are set in the piers to accept wood framing or posts.

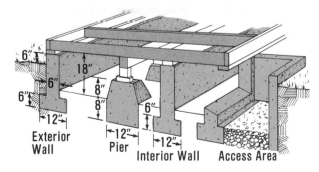

6"
18"
6"
8"
6"
8"
6"
12"
Exterior Wall
12"
Pier
12"
Interior Wall
Access Area

Example of pier and other common foundation combinations

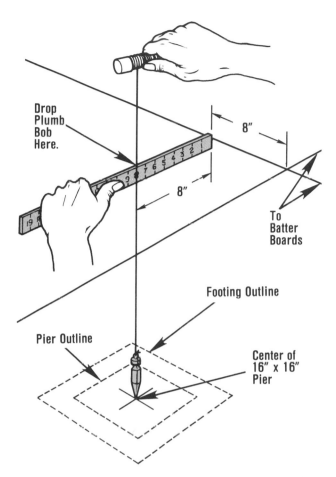

Drop Plumb Bob Here.

8"

8"

To Batter Boards

Footing Outline

Pier Outline

Center of 16" x 16" Pier

Locating footer and pier using batter boards and a plumb line

These simple foundations are commonly used for small structures with a crawl space design. Porches and wooden decks are two fine examples of structures that normally rely on pier or pillar foundations.

Footer and pier locations are found using batter boards, string lines, and a plumb bob as shown. Remember that the frame of the structure will be flush with the sides of the pier—a consideration to keep in mind when setting building dimensions and locating footer and pier locations.

Footings for pier or pillar foundations are always set below the frost level. Dimensions for footings and pier columns vary according to local soil conditions and the total weight to be supported. Firm, rocky soil, or bedrock, will bear much greater loads than soft clay, silt, or muddy soil. In areas of poor load-bearing soil, larger footings are needed to distribute the given load over a larger area.

Check with local building authorities for code recommendations in your area. As an example only, consider that a pier footing used to support a single-story vacation cabin or cottage might measure 3' square and be 15" thick. Footings for a pier foundation supporting a wooden frame porch or enclosed deck might be 2' on a side and 6" to 8" thick.

Footings must be absolutely level. In firm soil, an earth form can be used to pour the footing. The wet concrete will find its own true level. When the footing is larger than the pier, position and brace the form for the pier as shown. Pour the pier section after the footing is capable of bearing its weight.

When the soil is firm and capable of bearing heavy weight, the bottom of the pier can serve as its own footing as shown. Position the form so it rests on the bottom of the hole and make one pour. A simple pier form can also consist of a length of heavy cardboard tubing that is sold expressly for this purpose. Small precast pier foundations are also sold at building supply yards. These precast units often have post mounting hardware already in place. Both precast piers and tube-formed piers must rest on below ground footings.

In cases where the footer does not have to be large or deep and the pier or concrete pillar is fairly short, brace the pier formwork above the earth footer form as shown and make a single pour.

In all cases, the top of the pillar or pier form must be absolutely level. In addition, the piers must be set at the same height if the framing joists are to run between them. You must set these heights during form construction and positioning. A water level is the easiest method of ensuring that piers are set at the same height.

STEPS

Concrete steps and landings are attractive and durable and can provide good traction in

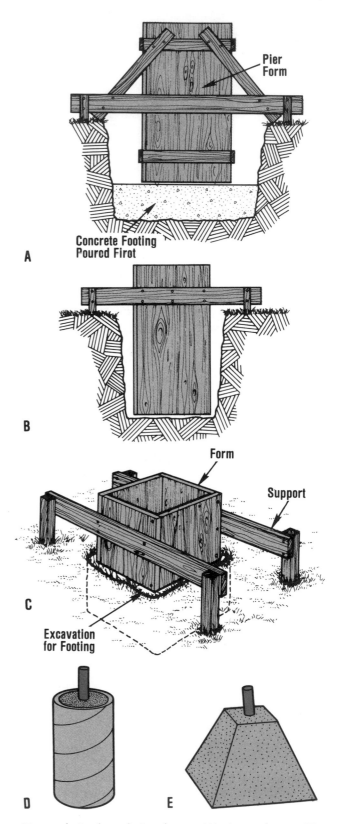

Types of pier foundation forms: (A) pier on footer, (B) footerless, (C) small earth footer and pier, (D) cardboard tube form, and (E) precast pier

Concrete steps with an embedded cobblestone finish. Oval cobblestone was embedded in partially set concrete using a procedure similar to exposed aggregate finishing.

When Riser Is:	Tread Should Be:
4″ to 4-1/2″	18″ to 19″
5″ to 5-1/2″	16″ to 17″
6″ to 6-1/2″	14″ to 15″
7″ to 7-1/2″	10″ to 11″

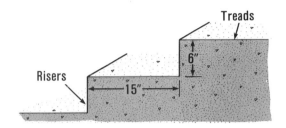

Riser–tread dimensions

wet weather. In most areas, steps at an entranceway to a home must conform to the provisions of the local building codes. These codes specify critical dimensions and design factors, such as:

- Width
- Height of flights without landings
- Size of landings
- Size of risers and treads
- Relationship between riser and tread size

Steps for private homes are usually 4′ wide and should be at least as wide as the door and walkway they serve. A landing is normally used to divide flights greater than 5′ in total rise. Landing size is usually no shorter than 3′ in the direction of travel. The landing at the entranceway may be 3′ to 5′ long and no more than 7-1/2″ below the door threshold. To assure proper drainage, landings and step treads should pitch downward 1/4″ per running foot.

Riser–Tread Relationship

The height of the riser and depth of the tread are important factors for achieving maximum safety and convenience. For flights less than 30″ high, maximum step rise is usually 7-1/2″ and minimum tread depth 11″. For higher flights, step rise may be limited to 6″ and tread depth 12″ maximum. Many studies have been undertaken to determine the finest riser–tread combination. One concludes that the sum of the riser and tread dimensions should always equal 17-1/2″ for optimum comfort and safety. This is a good rule of thumb for most steps whose main purpose is to transport traffic smoothly and quickly.

However, more generous tread dimensions can be used to create gently rising steps for patios, gardens, and terraces. Combinations such as 4:19, 4-1/2:18, 5:17, 5-1/2:16, and 6:15 are possible. As a general rule, the closer the climbing step comes to the normal walking stride, the safer and easier it is for all ages. Once you select a riser–tread ratio, use it for the entire project. Varying dimensions even slightly between steps can be very dangerous.

The use and design of footers for steps depends on the size of the project and the prevailing frost line. A small one- or two-step stoop built on a firm base in a mild climate might require no more than removing a thin layer of topsoil and tamping the base.

Stairs with more than two risers and treads can be quite heavy and should be supported on footers placed at least 2′ deep in firm, undisturbed soil and, in areas where freezing occurs, 6″ below the frost line. Steps should be tied to foundation walls with anchor bolts or tie-rods. On new construction, step footings should be cast as an integral part of the foundation wall.

On smaller projects, you can excavate the entire site to below the prevailing frost line, construct the formwork, and pour the concrete. This eliminates the need to cast a separate footer, but can also waste concrete if the project is large or the frost line deep.

On large projects in cold climates, the footer can consist of a 6″-thick concrete or concrete block wall set below the frost line. Concrete block must be set on its own concrete footer. A series of 6″- to 8″-diameter pier foundations can also be positioned at the perimeter of the project. Simply dig the holes below the frost line and fill them with concrete. When new steps are added to an existing building, this posthole method is an ideal way to prevent the new steps from settling front-first into the soil. Construct several posthole footings beneath the bottom tread and tie the top step or landing to the existing foundation with two or more metal ties or anchors.

Formwork

Two methods of finishing concrete steps are used, and each has a bearing on formwork construction.

The first method involves placing the concrete, finishing landing and tread surfaces, and then removing the riser forms and finishing riser surfaces as soon as possible thereafter. The two critical steps of this technique are timing the removal of the riser forms and working without damaging the relatively soft concrete. The time lapse can be anywhere from thirty minutes to several hours after placing the concrete, depending on the weather and setting conditions.

When this method is used, forms must be constructed for easy riser removal without sacrificing strength. This method allows complete finishing in one day and produces the finest, most consistent finish on risers and treads. However, this method might be too risky for inexperienced workers. Finishing the relatively soft concrete is difficult. Edges and surfaces are easy to damage and hard to repair. The work must also be completed fairly quickly, or riser surfaces will stiffen too much to be worked smooth.

For these reasons, many less-experienced workers obtain better results with the second finishing method. In this method, the concrete is placed and the landing and tread surfaces finished. However, the riser forms are left in place during the entire curing period. This eliminates the chances of damage and the rush to complete the work. After the riser and side wall forms are removed, any imperfections are repaired. With this method, formwork can be built without regard for fast, easy removal of the risers early in the finishing process.

It's very important that forms for steps are rigidly braced to prevent bulging and leaks. Forming lumber must be straight, true, and free from surface imperfections that will show in the surface of the concrete. Wood containing knots or blemishes will mar the surface finish and make finishing more difficult.

Step side walls can be formed with plywood or hardboard panels or with tightly fitted 2× planking. When using panels, riser forms must be cut to fit the inside dimension between panel sides. As shown in the illustration, dual side panels can be used to cast a step-wall combination.

Once the side walls are in position, install the riser form boards by beginning at the top and working your way down. This eliminates the need to stand on or lean against riser boards once you install them. Use a level to make certain the boards are not tilted or cocked side to side between the side walls. Set the top of each riser slightly below the bottom of the previous riser.

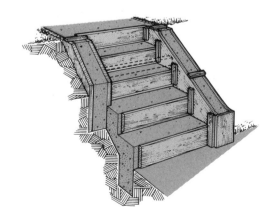

Step forms using double side walls. Risers are cut to fit between side walls and are attached with cleats.

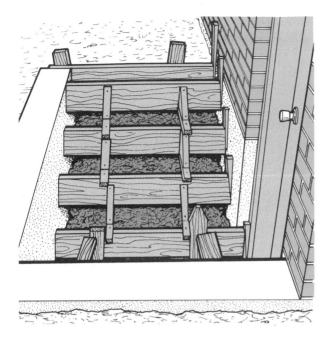

Step forms for stairway between existing walls or abutting one wall

This creates a forward slope on the treads of about 1/4", just enough to prevent puddling water.

Cutting an inward bevel on the bottom of the riser boards will enable you to finish the full width of the tread prior to removing the riser form. You can also tilt the entire riser board

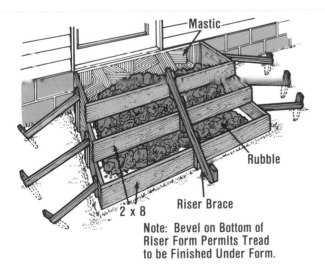

Note: Bevel on Bottom of Riser Form Permits Tread to be Finished Under Form.

Formwork for freestanding steps with landing. Center bracing is needed for all wide steps. Fill material should be clean and well packed.

inward an inch or so to increase the overall width of the tread at the base of the riser.

Riser boards must be securely fastened to side wall panels. Wood cleats are the best method of attaching risers to plywood forms. If the risers are being positioned between two existing masonry walls, cut them slightly short and use wooden wedges jammed between the risers and wall to hold them in place. In addition, wide runs must also be braced midway between the side wall panels and firmly staked to the ground outside the formwork.

Some type of isolation joint must also be installed where the top tread or landing of the new work meets the existing building. This is done to prevent bonding between the two surfaces, which is a condition that may lead to cracking in the future. A fiber strip can be used, but a thin layer of building paper positioned against the foundation or a coating of heavy asphalt mastic works just as well.

Finally, take the time to plan for the anchoring of any anticipated railings or ironwork on the steps or landing. This may involve forming recesses in the fresh concrete after placement using wood blocks or embedding actual hardware in the concrete.

Fill and Falsework

Very few, if any large step projects are solid concrete. Stone, old concrete, broken brick or block, and well-packed soil are normally placed inside the formwork to reduce the amount of concrete needed. Fill materials should be located no closer than 4" from the sides and top of the forms.

Inside forms, which is called *falsework*, can also be used to reduce the amount of concrete needed. Falsework made of wood or other absorbent material should be placed no closer than 6" from the face of any outside form. If the wood is covered with plastic sheeting or other moistureproof material, this distance can be reduced to 4". Poorly protected wood falsework will absorb moisture from the concrete and swell, generating internal pressures that can lead to severe cracking. To help prevent cracking in the top surface of a large landing, position iron rebar on top of the fill or falsework.

Placing Concrete

Use a 1:2-1/4:2-3/4 mix with a maximum coarse aggregate of 1" and slump not exceeding 3". If the forms are to be removed on the same day, simply wet down their inside surfaces; otherwise, apply a light coating of oil or release agent to the forms.

Begin placement at the bottom step and work upward. Fill in tight against the side wall and riser forms, working out any possible voids with a rod or shovel. Tapping the riser boards with a hammer will also help settle the concrete and bring the fines to the surface.

Float each tread as it is filled, being careful to create the needed forward slope. The top step or landing is filled last and then screeded level and darbied.

Finishing

Begin finishing operations at the top and work downward, one step at a time. Once the concrete loses its bleed water, edge along the landing and riser forms using a 1/4" or 1/2" radius tool. Follow with floating and troweling operations at the appropriate times. Draw a broom over the troweled treads to produce a lightly grooved, nonslip surface. If the risers are

Float finishing a riser face

to remain in place for the entire curing period, work is completed for the day, except for curing.

If you plan to remove the riser boards that same day, the toughest work is ahead of you. You must wait until the concrete has set up sufficiently to support its own weight before removing the riser boards. If the boards are removed too early the steps will sag and bulge outward, and no amount of work will return them to the desired shape. Once again, begin at the top and work down, one step at a time.

Immediately after removing a riser board, float the riser surface and use an inside step tool to dress the joint where the riser and tread meet. The radius of this inside finisher should be the same as the standard edger you are using. If the riser surface is slightly honeycombed or pitted and floating does not bring sufficient fines to the surface for smooth finishing, apply a thin mortar coat of one part cement to one and a half parts fine sand. Work this mortar into the honeycombed surface using a steel trowel. After troweling, broom finish the riser surface to match the treads. Work quickly and carefully. Too much time on any one step can cause the other to set too firmly for proper finishing.

The side forms can be removed the same day as placement or left in place for the entire curing period. If side walls are removed on the same day, the side surfaces should be floated and then plastered with a 1/8" to 1/4" layer of mortar applied with a steel trowel. This plaster coat can then be floated to produce a slightly rough finish or floated and troweled for a smooth finish. Brushed or swirl finishes can also be created after floating or troweling. (See chapter 6 for details.)

When the riser and side forms remain in position during the entire curing period, finishing work resumes on the day the forms are stripped. After the forms are removed, any small projections should be removed by chipping or rubbing the surface with a clean brick or abrasive stone.

Fill any honeycombed areas with mortar as just described. If surfaces are not uniform in color, saturate the affected areas with water and apply a light coat of mortar using a brush or rubber float. Vigorously float the mortar into the surface to fill any small voids. Scrape off any excess mortar and allow the surface to dry. Once dry, rub with clean burlap to remove any loose mortar. Surface color should now be uniform.

Keep all surfaces wet with water for the entire curing period. Pay particular attention to the side walls once their forms are removed.

PRECAST CONCRETE STEPS

Steps and porches of precast concrete have dramatically risen in popularity. These factory-made units greatly simplify construction. Precast steps with treads, risers, side walls, and porch all cast as one piece are available in a variety of styles and sizes. Widths of 3' to 6' with any number of risers up to 6 and porches up to 6' deep are commonly available from manufacturers of precast steps.

To meet the need for wide entryways, step units can be used in multiples. Also, separate precast porches are available in some areas in heights up to 42" and varying lengths and widths to meet the requirements of wide entryways.

Precast steps are used for new construction and as step replacements for older homes. The units are delivered by the manufacturer ready for installation. Often the manufacturer can install the units quickly with special trucks and hoists. Railings are bolted to inserts cast into the steps during manufacture.

There are several methods of installing precast steps. They may be bolted to the building's foundation wall and require no substantial ground support. An alternate method uses a simple slab foundation (often provided by the precast manufacturer). The steps may rest directly on the sidewalk, in which case it should extend all the way up to the building foundation. Deeper perimeter or posthole pier footings may also be used in cold climates. Consult the precast company about the method they prefer. The company will do the installation, but you might save money doing the site preparation and footer work yourself.

STEPPED WALKS AND RAMPS

Slight changes in elevation over moderate to long distances can be handled through the use of stepped walkways, which were discussed in chapter 5. A slight change in elevation over a short distance can be an ideal location for a ramp. Ramps also come in handy as paths for wheelbarrows, lawn mowers, and other equipment. Forms can be made of planking or plywood and must be firmly staked in position. Ramp thickness at its low end should be no less than 4".

Use a gradual, steady drop from the ramp's high end to its low end. The average thickness of the ramp will be the midpoint between the high and low thicknesses. For example, the amount of concrete needed to construct a 5'-long, 3'-wide ramp that is 4" thick at its low end and 8" thick at its high end is equal to the amount needed to construct a 6"-thick slab 5' long and 3' wide.

Place the concrete beginning at the low end. Use a slightly stiffer mix and allow the concrete to set up slightly so it helps support its own weight as the placement proceeds.

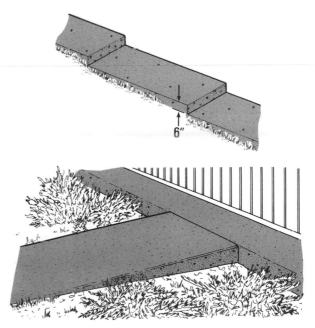

Handling slight changes in elevation can be done using stepped walkways or ramps.

Chapter 8

OTHER CONCRETE PROJECTS

Not all concrete work involves the placing of massive walls or wide expanses of slab work. Small concrete slabs provide a strong, level base for countless items inside and outside of the home.

Concrete's moldability makes it a surprisingly versatile and easy-to-work-with casting material for hobby-oriented projects. And its holding power makes it ideal for permanently setting posts and poles. But before we take a closer look at these and other unique concrete projects, one other major structural use for concrete should be covered.

CAST CONCRETE DECKS

A cast concrete deck is the most popular method of forming the landing area of a raised porch. The walls of the porch can be constructed of concrete or block (see chapters 7 and 15). Once the walls of the porch are built and proper-

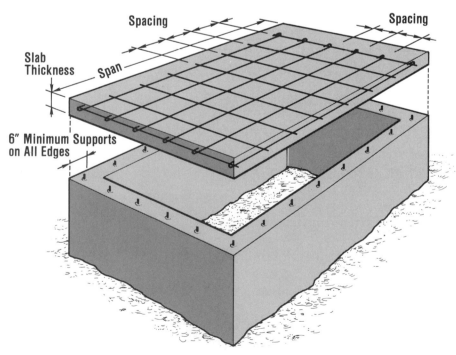

Reinforced concrete decks for porches made of poured concrete or block walls are supported on all four sides.

ly cured, you are faced with the task of capping the hole they created. The project is similar to the slab work that was covered in chapter 5, except you are working several feet off the ground with no existing subgrade.

Fill-in and Formwork

The easiest way to support wet concrete is to fill the hole with soil up to the top of the porch foundation walls. You should tamp the soil firm in layers as it is shoveled in. Old brick, block, stone, and concrete can also be added to the fill. (See page 245 for an illustration of this technique with a block porch.)

Wooden Bases. If fill material is in short supply, you may prefer to build a solidly supported wooden platform as the base for the deck concrete. The frame for the platform should be made of 2×4 lumber, spaced 2' on center. The face material should be 3/4" exterior grade plywood.

Make the platform slightly undersized for the hole, so when it is in final position there is a 1" gap between the edges of the platform and the porch walls on all sides. The platform itself is supported by up-ended concrete blocks. Position the blocks no more than 2' apart so that the

2×4s of the form will rest on them when the form is set in place. The form should be level with the top of the porch. Use mortar to seal the gap between the platform and the walls before pouring the concrete.

Side Formwork. Once the earth or platform base is ready, you can construct the formwork for the sides of the concrete deck. Deck thickness should be a minimum of 4". A deck thickness up to 6" is commonly used for added strength and durability—a factor you should consider. Check with your local building authorities for recommendations in your area. Since porch construction is considered a major home improvement, there are probably code specifications that apply to deck thickness and reinforcement.

Planks for the side forms can be 1" or 2" nominal thickness, and 8" to 12" wide to accommodate the overall deck thickness. One-inch thick lumber is lighter and, therefore, easier to support against the sides of the porch walls. The forms should overlap the sides of the walls 4" to 6" to provide adequate support. Forms must be firmly pressed against the sides of the wall and fully supported by stakes driven deeply into the ground. If the holes produced will not mar the finished wall, masonry nails can be driven

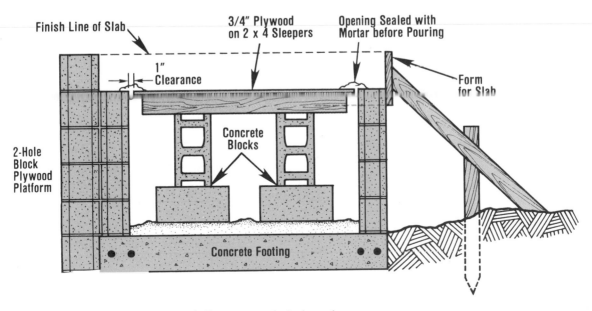

Formwork for concrete deck using a fully supported platform base

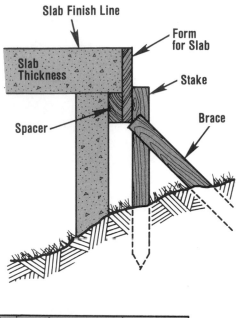

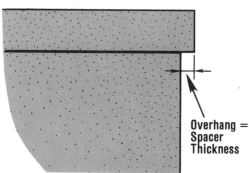

Details for forming an overhanging deck edge

through the formwork into the walls to help provide support. For easy removal, do not drive the nails all the way in. Never use masonry nails as the sole means of formwork support.

If you wish to create an overhanging slab, insert spacers between the porch foundation walls and the outside form boards as shown. The distance the form boards extend above the top edges of the walls determines the thickness of the slab. When setting the forms, pitch them 1/8" to 1/4" per foot to create a natural water drainage away from the house or structure. You must also install a suitable expansion or isolation joint where the concrete deck meets the existing structure.

Be certain all formwork is secure and the highest quality construction before proceeding

with the pour. This is particularly true if a wooden platform base is used. Any formwork failure during this project will be disastrous.

Steel Reinforcement

Because the slab will only be supported by its edges, the deck must be reinforced with steel rebar set at proper intervals. Unlike some types of slab work, this is not an option. The steel reinforcement is needed to prevent the deck from cracking when the fill or wooden platform beneath it begins to settle or decay. With a properly designed deck, the fill or platform is only a temporary support during the construction process. The steel rebar creates the permanent support and strength of the concrete deck.

The steel is positioned in checkerboard fashion about 1-1/2" from the bottom of the slab. The rebar is wired together to prevent it from shifting during the pour. It should also be tied into rebar protruding from the top of the concrete walls or grouted block walls. Rebar in concrete walls must be placed during wall construction. Rebar in hollow block walls can be placed during wall construction, but you can also position it and fill the top two courses with concrete during your deck placement. If you plan this method, be certain the hollow block cores are securely plugged with wire mesh two courses down from the top of the wall. The best time to place this mesh is during wall construction. Remember, if plugs fail during the deck placement, substantial amounts of concrete will be lost down the block cavities. You must also include the amount of concrete needed for grouting when estimating the overall quantity for the entire job.

Table 8–1 lists the various rebar sizes and spacings used on typical cast deck construction. You should always check these specifications against local codes.

Placement and Finishing

Use a high-quality 1:2-1/4:2-3/4 mix for the deck. A slightly stiff mix is best. Remember to plug the edges of the platform deck, if one is

A

B

C

D

Construction of a concrete deck for an enclosed sun porch built as part of a barn renovation project: (A) Deck walls were constructed of concrete block. Since the original grade was to the bottom of the barn stall doors, little additional fill was needed. Because of this stable base, the amount of rebar used was less than that recommended in table 8–1. (B) After firmly compacting the soil, a layer of rigid insulation board was placed over the fill and the rebar was positioned. During the pour, the top two courses of block were grouted with concrete. The strip of rigid insulation shown resting on edge was embedded in the concrete to help stop cold penetration through the slab from the outside. (C) To create a slightly overhanging lip, a wooden form was built along the edge of the wall. The form was lined with dark plastic sheeting to impart a smooth finish to the concrete. (D) The finished deck. A stone over block composite wall was built to extend the original barn stone foundation.

TABLE 8-1:	CAST CONCRETE DECK REINFORCEMENT SPECIFICATIONS*		
Deck Thickness	Rebar Size	Span	Spacing between Rebar
4″	3/8″	5′	18″
4″	3/8″	6′	14″
4″	3/8″	8′	10″
4″	1/2″	10′	10″
6″	3/8″	5′	20″
6″	3/8″	6′	18″
6″	3/8″	8′	14″
6″	1/2″	10′	12″

* Common rebar sizes and spacings used on typical residential loads of 50 lb/sq. ft. Check and follow local code requirements.

Note: Data based on decks supported by walls on all four sides. If deck is not square, use longest dimensions in calculations.

being used. It's also advisable to do any block grouting before starting the deck pour full force.

Place the concrete very slowly and carefully, avoiding any sudden weight shocks to the formwork. Tap the sides of the form to help the concrete settle and fill all voids. When the forms are completely full, screed, float, and finish the concrete as you would any other slab, with the exception of control joints which are not needed in highly reinforced concrete decks.

If you plan to use the deck as a base for flagstone or other paving materials, screeding, tamping, and bull floating are all that is required. Provision for railings and other items to be attached to the concrete deck should be made during the placement. Center load-bearing hardware over the porch walls. If the exact location of the item can be determined, embed the appropriate mounting hardware in the fresh concrete. You can also form mounting holes by gently embedding wood blocks about 2″ square by 4″ long. Remove the blocks after the concrete begins to set up, but before it reaches final hardness. A nail driven into the top of the block will make lifting out easier. Hardware can then be mounted in the hole using anchoring cement at a later date.

CAST CONCRETE PROJECTS

While concrete lacks the plasticity to be hand sculpted like clays or putties, it is fluid enough to be shaped by more intricate molds than the heavy lumber forms used in structural construction. Planters, benches, pavers, birdbaths, posts, splash guards, anchors, bookends, decorative plaques, cold frames, boat docks, and chimney caps are some of the ideas presented in this section. The principles discussed here can be used for other projects as well.

Mixes for Cast Projects

The mix you choose for your project will depend on its function and shape. (All mix proportions given in this section are by volume.) Items such as benches, pavers, and the walls of a cold frame or hotbed will need a standard high-strength 1:2-1/4:2-3/4 mix. The size of the coarse aggregate should be 1/2″ or less to help make smoother surfaces possible. Decorative stone aggregate is commonly used to create more elaborate surfaces on planters, pavers, and the like.

When load-bearing strength is not needed as much, substitute a sand mix that can be purchased premixed with one part portland cement to three parts clean sand. The sand mix should be strengthened with an acrylic fortifier. Wire or thin rod reinforcement may also be needed. Special surface-bonding cements sold for mortarless block construction or surface renovation work also make good casting materials. Their fiber-based reinforcement gives them good flexibility.

For items that only need to support their own weight, you can use special lightweight mixes that substitute vermiculite, pumice, crushed lava rock, perlite, or haylite for the standard sand and gravel aggregates.

Vermiculite is used by gardeners to aerate tightly packed soils and by builders as an insulating or soundproofing material. Garden vermiculite is finer and more tightly packed. This produces a denser concrete that is slightly easier to shape and carve. Although vermiculite produces the lightest concrete possible, it has low-strength and little or no resistance to weathering. Use vermiculite concrete for indoor or sheltered projects only. The lightest possible mix is produced by mixing one part cement with four to six parts vermiculite. A slightly stronger 1:2:3 mix of cement, fine sand, and vermiculite is also popular. When molds contain a good amount of fine detail, increase the cement content slightly.

Pumice, crushed lava rock, perlite, and haylite are normally substituted for the coarse aggregate in 1:2:3 mix ratios. These mixes are somewhat heavier than vermiculite concrete, but they are stronger and can be used outdoors.

No-fines concrete combines cement and 3/8" maximum coarse aggregate, while omitting the sand or fine aggregate. A 1:9 mix by volume is a good starting point. Use just enough water to create a paste that acts like glue. It should flow and coat all surfaces. Add water a little at a time until you reach this consistency. When molded, this mix produces a deliberately honeycombed or open surface that is excellent for all nonload-bearing items.

Craft concrete projects are an area open to experimentation and a little fun. Cast stone can be made using marble dust and finely crushed marble as aggregates. Maximum aggregate size for small projects such as decorative plaques or bookends should be no more than 1/8". You can also increase the cement content of smaller projects for increased strength and workability. Ratios of one part cement to one part total aggregate have been used successfully.

Other unusual aggregates include glass beads, wood chips or sawdust, and crushed nut shells. White portland cement can be used to brighten and highlight the final finish, or colored pigments can be added to the mix. Trial patches and test blocks are almost a must for some of the more exotic materials mentioned, and always stick to standard mix ratios when the project must bear weight or function structurally.

Formwork Materials

Cast projects can use a wide range of form materials. In addition to commercial aluminum molds, cans, buckets, cardboard boxes, mailing tubes, and automobile tires can be used to create useful indoor and outdoor accessories. You can also create interesting effects, depending on the specific form material you use. For example, unplaned or resawed lumber leaves a woodgrain effect on the concrete, while lining the form with heavy kraft paper or plastic produces a smooth finish.

Make forms easy to remove to minimize the risk of damaging the cast item. Use duplex-(double-headed) nails and always apply a light coat of oil or release agent to interior surfaces.

Direct relief molds for pavers and decorative slab work can be made using wet or oiled sand, wood, clay, or rigid foam. If the mold or form is being used to create the final shape, give the cast project plenty of time to cure before removing it. Cure the piece on a raised platform to stimulate air circulation on all sides. Wrap it with wet burlap or cloth and allow five to seven days for curing.

In highly decorative work, the mold may create the initial shape only. Once it is carefully removed, the piece can be shaped further by carving with a sharp knife or other tool. Lightweight vermiculite mixes are ideal for this mold and carve technique. Formwork may also be removed a little early when working with decorative surface aggregates so the surface can be brushed clean.

BOOKENDS

This is a good starter project. Because it is small, you can experiment with various lightweight concrete formulas and decorative finishes without wasting any substantial amount of materials:

1. Place a tight-fitting wood divider into a small or medium-size coffee can.

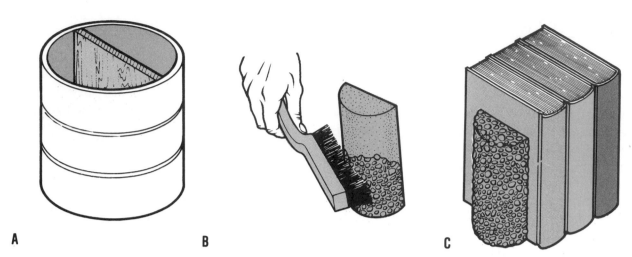

A   B   C

Simple lightweight concrete bookends are made with a coffee can mold.

2. Pour lightweight concrete or special aggregate mix into the mold so it is filled completely.
3. Smooth the top of the mold level.
4. When the concrete has set up and cured sufficiently, cut away the can mold using metal snips.
5. If lightweight concrete was used, the bookends can be further enhanced by carving relief patterns and decorative features.
6. If special aggregates were used, scrub and buff the surface.
7. Finally, glue a piece of heavy felt to the bottom of each bookend to avoid scratching shelves and furniture.

ANCHORS

An anchor was never this easy to make:

1. Start this project with an old bucket, a large can, or a similar container.
2. Pour regular concrete, sand mix, or surface-bonding cement into the container until it is completely full. After about twenty to thirty minutes, place an anchor bolt in the container as shown. Trowel the surface smooth.

With anchors, embed the eye bolt or hardware deep inside the body of the concrete.

3. Do not attempt to use the anchor until the concrete has cured completely. If additional weight is needed, make several anchors and join them with chain.

WHEEL CHOCKS

This project is perfect for the "Saturday mechanic"—wheel chocks connected by clothesline for easy transporting.

1. Start with two cardboard boxes approximately 8"×12"×6". Remove one of the small sides from each box. Next, cut diagonally with a utility knife to remove half of each rectangular side as shown.

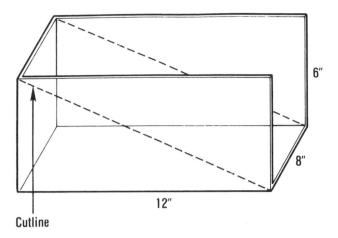

Cutline

Formwork for simple wheel chocks can be made from sturdy cardboard boxes. If the box has a top and bottom, only one is needed. Apply duct tape to seams and corners for reinforcement.

Anchor Bolt

Wheel chock in use

The forms for the wheel chocks are now complete.

2. Use masking tape to fortify the 6" side of the forms, then prop up the forms 3" to 4" with wood blocks as shown.
3. With the utility knife, puncture a small hole in the center of the 6" sides. Insert an anchor bolt into the boxes; when the concrete hardens, the bolts will become firmly set in the wheel chocks.
4. Pour the standard concrete mix into the boxes so that they are completely filled.
5. When the concrete has partially hardened (about twenty to thirty minutes), smooth it with a trowel, then take the forms off the wood blocks.

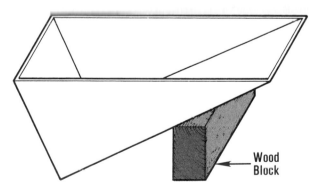

Wood Block

Position form for filling with concrete.

6. When the concrete has cured completely, rip away the cardboard boxes. As an easy way of carrying both wheel chocks at once, tie each end of a short length of rope around the anchor bolts and use it as a handle.

DOWNSPOUT SPLASH GUARDS

A concrete downspout splash guard is durable, long-lasting, and easy to build with these step-by-step directions.

1. Dig a trench the required length; it should be approximately 12" wide and 6" deep.
2. Use 1×4 lumber to build the forms. Stake and nail boards firmly in place, then apply a thin coating of engine oil to the form.
3. Pour concrete, sand mix, or surface-bonding cement mix into the form. Use a trowel to shape the gutter as shown; it should be at least 4" deep at its deepest point. Allow at least one week curing time before using.

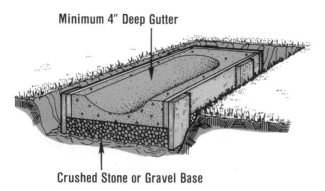

Downspout splash guard

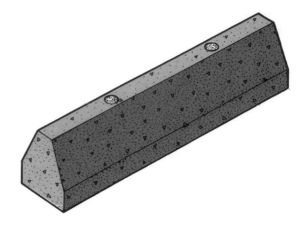

Finished cast curbing

4. Be certain to always slope the gutter away from the downspout in order to maintain proper drainage.

GARDEN EDGING/CURBING

To accentuate your home's exterior, try this project—garden edging/curbing.

1. Construct a "V"-shaped form from 1×2s; a good workable length is 4', although this can vary.
2. As shown in the cross section, the form should be approximately 10" high and at least 4" wide at its widest point, which will eventually be the base of the edging. Making the base any narrower will cause the edging/curbing to be too unstable.

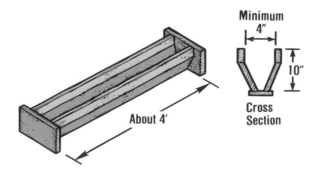

Forms for curbing and edging

3. Apply a thin coating of engine oil to the form. Pour standard concrete, sand mix, or surface-bonding cement mix into the form until it is completely filled. Trowel the surface smooth, then allow at least forty-eight hours curing time before removing it from the form.
4. If you want to stake the edging/curbing into the ground, simply cast the project with holes in the center as shown. Properly positioned and sized dowels can be used to create the holes in the fresh mix. Pull them out before the concrete sets completely.

PLANTERS

Planters are a good project for exposed aggregate finishes, white or colored mixes, and lightweight concrete. The thin cross-section of the walls limits the size of the aggregate used. The planter walls must also have some rigidity and strength, so lightweight concretes should be proportioned with some load-bearing aggregate.

Rectangular Wooden Forms

1. Typical outside dimensions of a cast planter might be 18" long × 15" wide × 10" high. The side walls of the planter form are made of 3/4" thick plywood or

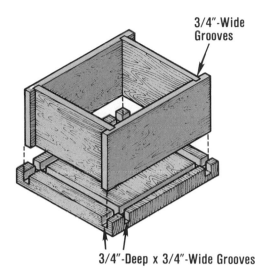

3/4"-Wide Grooves

3/4"-Deep x 3/4"-Wide Grooves

Forms for rectangular planter outer walls

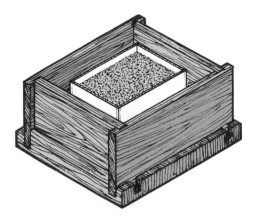

Forms for rectangular planter inner walls

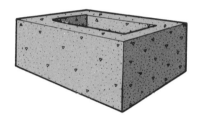

Rectangular planter

planking. If the grooved assembly method shown is used, the base lumber must be at least 1-1/2" thick. If the form will be nailed together for one time use, a thinner base can be used.

2. Assemble the form by means of a grooved system illustrated or nail the pieces together, cleanly butting adjoining sections together.

3. Drill a 1/2" hole through the center of the bottom, then plug it with an oiled dowel that is long enough to penetrate the casting. The dowel will act as a drain hole for the casting.

4. Prepare the inner form by filling a cardboard box, a large can, or even a wastebasket with sand. For a double-cavity planter, use two containers. Do not put the inner form in place until after the concrete base has been poured.

5. To make removing the form easier, coat the inside with a thin layer of oil.

6. To cast the planter, pour the base layer of concrete, add the inner form, then fill in around it. Use a tamper to firm the concrete around the form as the pouring progresses.

7. When the concrete has set, remove the form. Clean the form parts immediately after removing them, then set them

aside for future use. (They must always be reoiled before each use.) Allow the concrete to cure three to four days before using the planter.

Sand Cast Round Forms

As you might expect, casting a round planter requires a slightly different approach. In this case, the outer mold is made using a bed of damp sand. The sand can also be oiled to compact it if there is a problem with drying out during the casting process. Once again, lightweight or decorative mixes with sufficient strength should be used.

1. Scoop out an approximate-sized hole in the sand, then use a wastebasket, drum, or similar container to complete the cavity.

2. Compact the sand around the container. As you bear down on the container, twist it to make a good impression.

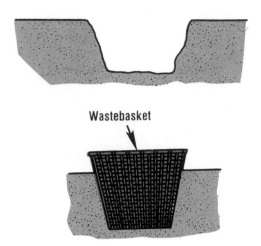

Creating a round outer form using sand compacted with water or clean motor oil

When removing the container, twist it as you lift.

3. Pour some of the mix to form the base, then insert a greased dowel to form the drain hole as shown. Make sure the dowel is long enough to penetrate the casting.

4. Choose a container for the inner form. It should be similar to the container used to form the cavity, but smaller; a square

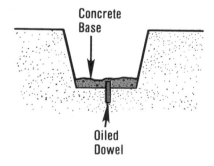

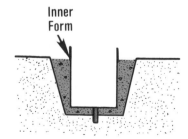

When placing the planter base, a dowel forms the drain hole.

Finished round planter

box or some type of cylinder is a good choice as long as it provides adequate wall thickness.

5. When the inner form is in place, pour the remaining concrete around it. Use a tamper to form the concrete around the form while you pour.

6. Wet-cure the planter in the sand for three to four days.

CAST-YOUR-OWN PAVING

When designing a walkway or patio, consider casting your own pavers instead of buying them or using brick or stone. You can even do the casting indoors over the winter months and stockpile the pavers for a spring project. Just be sure to work in an area where the curing pavers will not freeze. Many interesting geometric or interlocking designs can be created using simple formwork. Use standard concrete proportions

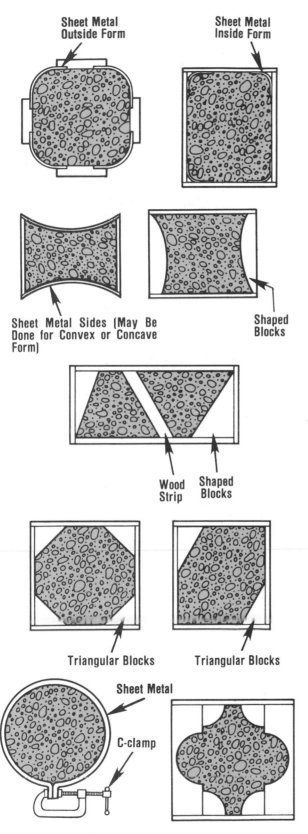

Sheet Metal Outside Form

Sheet Metal Inside Form

Sheet Metal Sides (May Be Done for Convex or Concave Form)

Shaped Blocks

Wood Strip

Shaped Blocks

Triangular Blocks

Triangular Blocks

Sheet Metal

C-clamp

Ideas for paver formwork

for the mix, although you may wish to limit the coarse aggregate size. The simplest method of casting pavers is to dig an earth form in firmly packed soil. Fill the hole with the prepared mix, smooth the surface of the pour, and allow it to cure fully. The paver can then be removed for use. Hinged, reusable wooden forms are also easy to build and use.

For special finishes, you can seed the bottom of the mold with a layer of decorative aggregate before pouring in the concrete. Pour slowly as not to displace the decorative stone. After curing, this exposed aggregate becomes the top surface of the paver. You can also place wood strips, leaves, and other shallow objects in the form to create interesting relief patterns in

Soil Form for Homemade Pavers

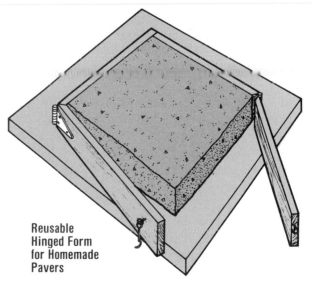

Reusable Hinged Form for Homemade Pavers

Earth and wooden forms for cast pavers

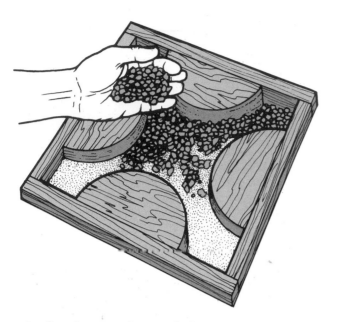

Seeding the paver form with decorative aggregate

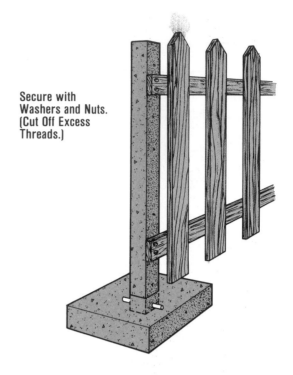

Secure with
Washers and Nuts.
(Cut Off Excess
Threads.)

the paving face. If you are using coloring pigment, establish a precise formula for all ingredients, including water. Maintain this formula throughout the entire project for proper color matching.

CAST POSTS

Small aggregate concrete, sand mix, or surface-bonding cement mixes can be cast into easy-to-make forms to create a variety of post designs with comparatively small cross sections. Cast posts can have their own bases or be set in concrete.

1. Make wooden forms carefully, using a minimum of parts so stripping will be easy. You can use wood moldings to create more stylish posts.
2. Coat all concrete–contact surfaces with oil or release agent.
3. Install 1/2" reinforcement rods when making solid posts. Install conduit or pipe in posts that will contain electrical cables.
4. Pour the mix into the forms and level it with a square trowel or straightedge. Fill the form completely.

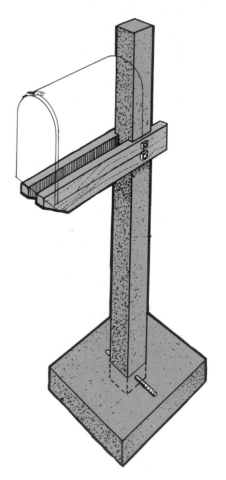

Cast post using integral base

Shaping Cast Posts.

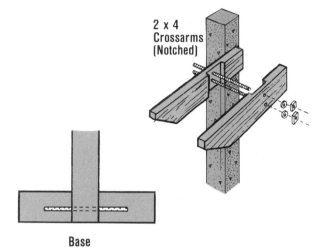

2 x 4
Crossarms
(Notched)

Base

Wooden Forms

Portable Posts

Portable posts have a number of uses, including clothesline posts, tetherball posts, and supports for badminton or volleyball nets. They are easily made with a heavy-gauge steel pole, a pipe sleeve (with an inside diameter equal to the outside diameter of the pole), a 5-gallon drum, and a base made from a concrete mix.

1. Sink a 5-gallon drum into the ground, so that its top surface is just below grade. Be sure that the base it rests on is firm and level. (If a 5-gallon drum is not available, you can set the pole and sleeve directly into a concrete-filled hole in the ground.)
2. Position the pipe sleeve and pole in the center of the drum, then fill the drum with concrete mix. Make sure that the pole is plumb.

Another approach utilizes a discarded automobile tire as a permanent, above-ground

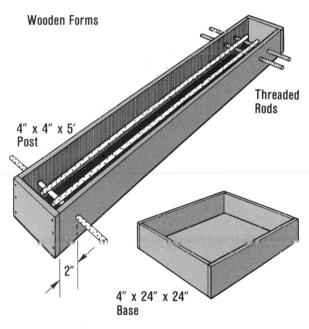

4" x 4" x 5'
Post

Threaded
Rods

2"

4" x 24" x 24"
Base

Forms for typical cast posts

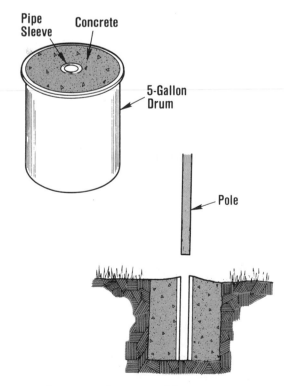

Pipe
Sleeve   Concrete

5-Gallon
Drum

Pole

5. Install any brackets, bolts, studs, hinges, light fixtures, hangers, etc., before the concrete has set.
6. Let the casting stand for at least forty-eight hours before removing the forms. Moist-cure for five to seven days.

In-ground mooring for a movable post

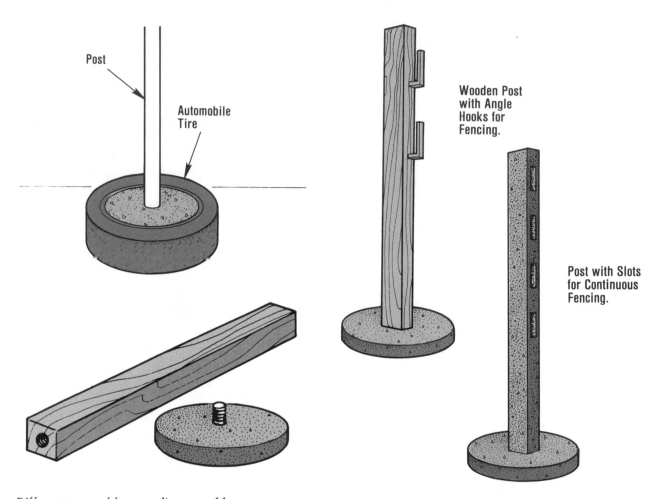

Post

Automobile
Tire

Wooden Post
with Angle
Hooks for
Fencing.

Post with Slots
for Continuous
Fencing.

Different types of freestanding portable posts

form. This design supplies ample rigidity and can be easily moved by tilting the post and rolling it about on the tire.

1. Place the tire on firm, level ground or on a sheet of plywood.
2. Center the post or sleeve and pour concrete mix around it.
3. Tamp enough concrete to fill the tire completely. Use enough concrete mix to slope the top surface so that water will drain off.

A post made of wood or cast out of surface-bonding cement, mounted on a base made of surface-bonding cement, has numerous uses as temporary fencing, including party lawn dividers, crowd-control barricades, hurdles, and horse jumps. Surface-bonding cement weighs as much as concrete. But its fiber reinforcing makes it less brittle, stronger, and more malleable than concrete so it can be cast in smaller dimensions. A bolt set in the base allows disassembly for easy storage.

BIRDBATHS

The birdbath shown is a slightly more involved casting project having three separate parts: the bowl, pedestal, and base. The most difficult piece to cast is the bowl, which is cast bottom side up over a clay or oiled sand core, disk, or similar convex surface. This convex surface shapes the inside of the bowl and a bowl template is used to shape the outside of the bowl. The casting material should be small aggregate

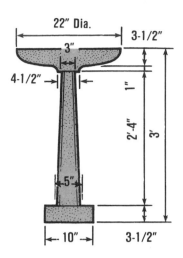

Dimensions for a birdbath cast in three parts

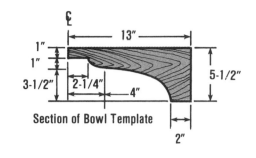

Section of Bowl Template

A

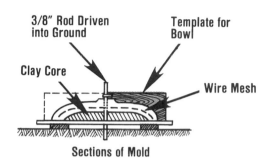

B   **Sections of Mold**

Bowl details: (A) template and (B) construction

concrete, sand mix, surface-bonding cement, or general purpose lightweight concrete.

1. Find or make a core or disk with an 18″ diameter and a 1-1/2″ thickness at its center. If you must make the core, construct a core template with the dimensions shown. Attach a leather or sheet metal strap to the end of the template. This strap is used to anchor the template to a center pin as the template is moved in a circle around the pin (a 3/8″ rod). Shape a clay or oiled sand core on a

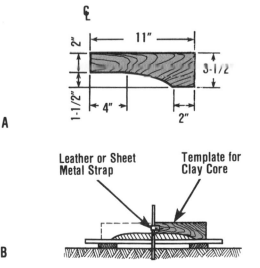

A

Leather or Sheet Metal Strap   Template for Clay Core

B

Core details: (A) template and (B) construction

wood base with a 3/8″ rod in its center and the template anchored to the rod, as shown.

2. Make a bowl template with the dimensions shown and attach a leather or sheet metal strap to its end.

3. When the core is dried, apply a light coat of oil and then place a low-slump concrete mix over it to a depth of about 1″.

4. Place wire mesh reinforcement over the layer of cement, then add another layer of 1″ thick concrete, and shape it with the bowl template.

5. Remove the 3/8″ rod from the center and set a 1/2″×3″ bolt in the exact center with the threaded end projecting out about 2″. Then cure the concrete.

6. For the pedestal, make two boards with a 2′-4″ length, a 5″ base, and a 4-1/2″ crown, and two boards with a 2′-4″ length, a 7-1/2″ base, and a 6″ crown. Nail them together, as shown, to make a form. Place 3/4″ molding in the four corners of the form.

7. Build a form for the base with the dimensions 10″×10″×3-1/2″. As an alterna-

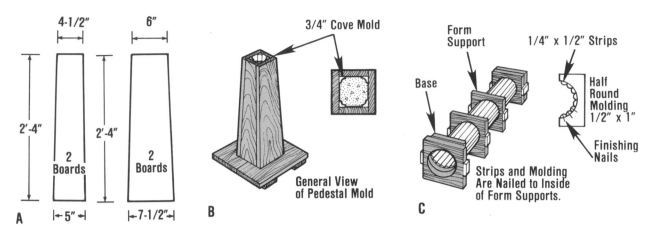

Pedestal details: (A) dimensions, (B) flat-sided forms, and (C) alternate rounded strip molding forms

tive, you can set the pedestal form over a posthole dug below the frost line.

8. Pour the concrete into the forms for the pedestal and base.

9. Center and set a 5/8″ greased dowel 3″ deep into the top of the pedestal. The dowel is removed after the concrete sets; it creates a hole to receive the bolt projecting from the bowl.

10. After the parts have been covered and cured, remove the forms, place the pedestal on the base and attach the bowl to the pedestal.

BENCHES AND TABLES

Benches and tables help to make outdoor living areas more relaxing and inviting. Because there is a large variety of designs, you can choose something particularly suited to enhance your own private retreat. You can cast tables and benches all in one piece, or you can make the slabs and the legs separately and then join them together. And they don't have to be made entirely of concrete either. You can top concrete legs with a wooden slab or use a concrete slab on wooden or wrought iron legs. You may want to combine concrete with brick, block, or stone instead. Whatever your choice, always use standard load-bearing concrete mixes.

For a one-piece bench or table, all inside dimensions of the form must be equal to all out-

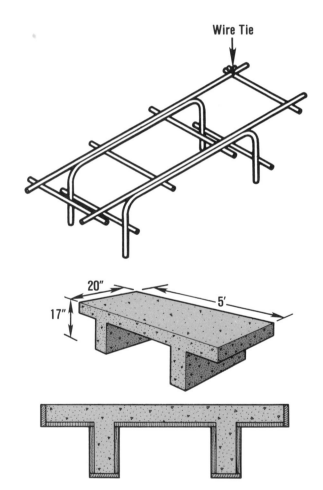

Forms and reinforcement for a one-piece cast bench

side dimensions of the finished project. The form must be sturdy and it should be assembled with butt joints so that it can be stripped away easily after the concrete has cured.

A one-piece concrete structure of this type also requires a skeleton made of 1/4" or 3/8" reinforcement rods tied together with wire. This skeleton should be constructed so that it will sit midway in the pour in each section. Make sure that the rod ends do not come any closer than 1" to the surfaces of the project.

After you have made the form and the skeleton for the project, follow these steps:

1. Set the form on a level surface, preferably in the shade, and coat all concrete-contact surfaces with oil.
2. Start filling in the legs with concrete and tamp the concrete to settle it in the form.
3. Set the skeleton in place when the legs are a quarter full. You can keep it elevated in the slab area by using a few small pieces of stone as temporary platforms.
4. Continue to pour the concrete. Be careful not to knock the reinforcement out of position. Continue tamping the concrete as well, and tap frequently on the outside of the form with a hammer.
5. Once the form is full, screed the top. Finish it off with a float. Use an edging tool to get round corners on the surface.

6. Allow the concrete to cure for three or four days. Cover the project with any material that will help to retain moisture. After the second day, very gently turn the project onto its side and then turn it so that its legs are sticking up like stalagmites. Let the project sit in this position for another day or two.
7. Remove the form. You can try to remove it in one piece by having one person at each end tug gently upward. If it does not move rather easily, however, you will have to disassemble it.

You can also make a table or bench in sections. The plans for one such bench are shown here. Use straight, dressed 2×4s for forming and three 3/8" reinforcing bars wired crosswise to three reinforcing bars 16" long for the seat section. Four 1/2"×3" bolts are set in the concrete of the seat section to anchor it to the legs. Four 5/8"×3" bolts are used in the leg forms to create sockets for the bolts in the seat section.

Set the oiled forms on an oiled wood base or on a concrete slab that has been covered with a polyethylene sheet. Place a 2" layer of concrete in the seat form and work it carefully along the inside edge of the form with a trowel so that it is tight against the form. Level the concrete and set

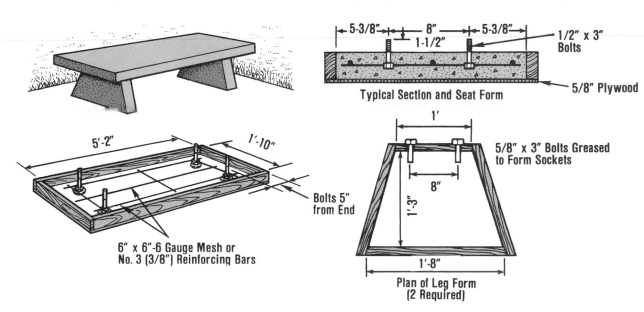

Forms for casting a simple bench in three parts

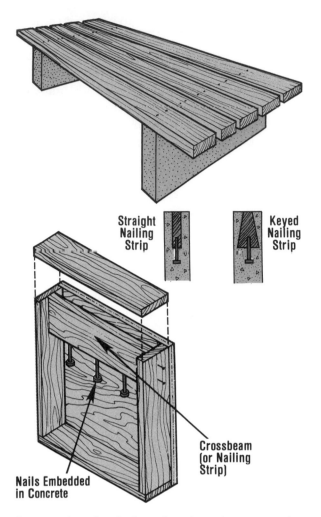

Straight Nailing Strip

Keyed Nailing Strip

Crossbeam (or Nailing Strip)

Nails Embedded in Concrete

Construction details for a bench with concrete legs and a wooden seat

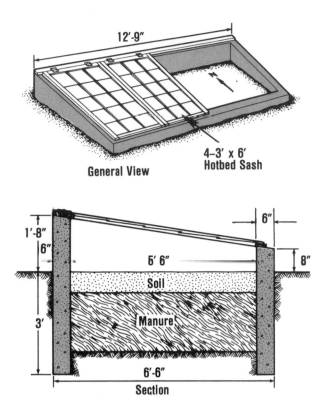

12'-9"

4–3' x 6' Hotbed Sash

General View

1'-8"
6"
6"
6' 6"
8"
Soil
3'
Manure
6'-6"

Section

Hotbed made with cast-in-place walls. Formwork is similar to that illustrated in chapter 7 for wall construction. Footings are not needed.

the reinforcing bars in place. Fill the form with concrete and strike it off level with a 2×4. Smooth the surface with a wood float. Once the mixture stiffens and the watery sheen is gone, use a steel trowel for the finish. Use an edging tool for smooth, dense, well-rounded edges. Fill and finish the leg forms in the same way.

If you want to build a bench with concrete legs and a wooden seat, use the form shown as a model. The cross beam of the form becomes part of the pour and is used as a nailing strip when the rest of the form is removed. It can be a straight board or keyed to provide a better joint. Note that galvanized nails (16d) are used to anchor the nailer in the concrete.

COLD FRAMES AND HOTBEDS

You can grow garden flowers and other plants in your own hotbed or cold frame. The two structures are very similar. The difference is that hotbed walls usually extend about 3' below the ground level to hold a deep layer of manure. In either case, a southern exposure to full sunlight is recommended. Try to provide protection from the wind as well as good drainage.

Simple concrete construction with precast panels or cast-in-place walls and floors makes a lasting structure that will be free from decay and insect damage. Make the pits in multiples of 3', which is the width of a standard hotbed sash. For a four-sash bed, the outside dimension would be 6'-6" × 12'-9". You can check with a local supplier for available sizes of storm sash or barn and utility sash.

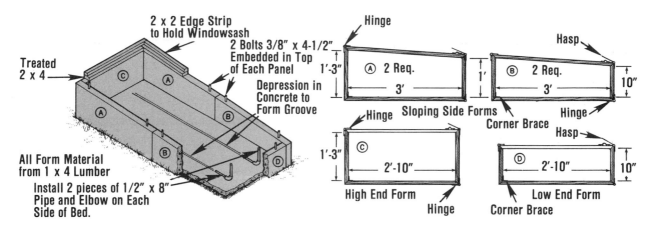

Formwork details for a cold frame made of precast panels cast flat and then set on edge to create walls

The sides of a hotbed or cold frame may be precast in forms like those shown. Wood strips are anchored in the top edge of the frame to permit hinging of the sash. Subsurface watering can be accomplished economically by making grooves in the concrete bottom with a piece of 2″ pipe. A 1/2″ elbow is set into the concrete with the open end directed toward the groove, and an 8″ length of 1/2″ pipe is screwed into the upright opening of the elbow. A funnel is used to pour water into the top of the pipe.

BOAT LAUNCHING RAMPS

You don't have to build a long solid ramp to allow for greatly fluctuating water levels; you can use a series of concrete planks linked together by steel cables. These planks make an economical, yet movable boat launching ramp. They are adaptable and will fit the existing contours of the lake shore or river bank. However, the planks should be used over sand or on a base that provides uniform bearing. The form used to make these planks is shown here. The cables hold the planks in place and serve as a guide for backing up cars. Apply a broom finish to the surface of the planks to provide good traction for cars backing into the water.

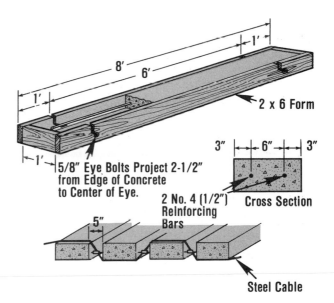

Precast concrete planks can be used to construct an economical and movable boat launching ramp.

CAST CHIMNEY CAPS

The top of all chimneys should be protected by a concrete cap that prevents water from running down next to the flue liner or puddling and freezing on the top surface of the brick or block. A simple sloped bed of mortar or concrete can be used, but more elaborate cast caps make an interesting weekend project.

The wooden formwork can be constructed on the ground and hoisted to the roof. The inside

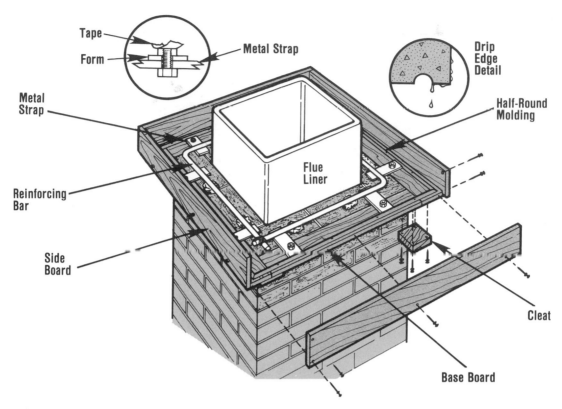

Formwork for a cast-in-place chimney cap

opening of the form should be sized to the outside dimensions of the chimney walls. Metal straps bolted to the base board pieces overlap onto the top chimney surface to hold the form in position. Bolts should be installed from the underside, with the securing nut located inside the form. When the chimney has set and cured, the bolts can be turned out and the base boards removed. To prevent wet concrete from clogging the threads of the nut or bolt, size the bolt length carefully so it does not protrude from the nut and also seal the top of the nut with a piece of tape before placing the concrete. The straps can remain in the cap or can be pried out after the forms are removed. If you plan to leave them in place, use strap material that will not rust and stain the chimney over time.

Use cleat blocks across the simple butted corner joints to increase strength at these points. The cap should also be reinforced with rebar. Lengths of curved strip molding positioned in the form, mitered at the corners, and tacked

secure will produce a drip control edging along the base of the cap.

Mix the concrete or sand mix on the ground and pass it up to the roof one bucket at a time. Layer the concrete on all sides of the flue to help stabilize the position of the form. Loading one side of the form while the opposite side is empty will tip the form upward.

Once the form is full, allow the concrete to stiffen slightly before beginning to build up the slope needed for drainage away from the flue liner. Use a wooden or magnesium float to build up the slope, working from the outside edge of the form to the flue wall. See the illustration shown on page 255 in chapter 16. You should also form a small radius curve along the edges of the form using an edging tool.

After finishing is complete, wet down the surfaces of the cap and cover it with plastic sheeting to prevent water loss. The forms can be carefully removed after one or two days. Curing should continue for five to seven days.

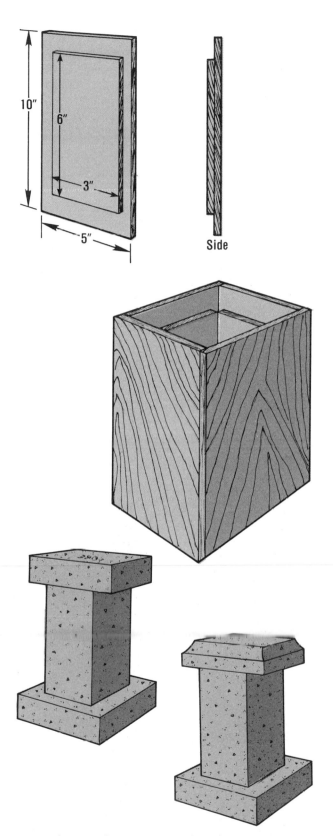

PROPERTY MARKERS

A boundary dispute is one of the most vexing situations that can confront a property owner. Many disputes are the result of survey features having changed over the years, pins that have been partially covered or buried by overgrown weeds, or stakes that have been removed because they were not recognized as survey markers.

Cast concrete markers are solid, durable, noticeable and, because they are clearly identifiable as markers, unlikely to be accidentally removed. You can make your own following this procedure:

1. Cut a plywood sheet into eight sections with the following dimensions: 5"×10" (2); 4"×10" (2); 4"×6" (2); and 3"×6" (2).
2. Center the 3"×6" pieces on the 5"×10"s and the 4"×6" pieces on the 4"×10"s. Glue or tack the pieces together to make sides for the form.
3. Coat the interiors with a thin film of oil.
4. Fit the sides together and bind them at the top and bottom with the wire.
5. Fill the form with concrete. Poke a stick, broom handle, or similar object into the form to compact the concrete and remove any air pockets. Screed the concrete flush with the form.
6. After the concrete has received its initial set, customize the marker cap with the date of installation, survey, and the owner's name.
7. Allow the concrete to cure for at least a day before removing the form.

DECORATIVE PLAQUES

Decorative plaques with raised or recessed surfaces can be created just for fun or as flower bed edging or garden accents. When used as edging, the plaques are partially buried to stabilize their position, but no separate footer or anchoring is needed.

Forms for simple property markers

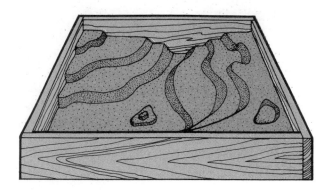

Direct relief mold using compacted sand set in a wooden box form

The form illustrated is a direct relief mold made of damp or oiled sand resting on a sheet of plywood. Wooden strips form the sides of the mold. The compactible sand is placed in the form and shaped using a variety of sculpting tools. The relief mold forms a negative image of the final textured surface of the plaque. Strips of wood or rigid foam can also be placed in the form to create recessed shapes or even lettering.

For raised or embossed shapes and letters, openings must be cut in the base of the form. To create a level surface for the embossed areas, the form should be placed on another plywood sheet or a screeded bed of compacted sand.

Concrete for plaques should consist of a sand mix, surface-bonding cement, or lightweight blend. Large aggregate mixes cannot be cast to create fine details. Place the mix slowly and carefully so as not to disturb the sand casting or move the rigid foam or wood strips. To help hold them in place, foam or wood strips can be glued to the base of the form.

Allow the mix to set up for at least a day before carefully removing the forms. With vermiculite-based concretes, additional carving can be done. Allow the piece to cure for five to seven days.

CAST CONCRETE POOLS

A simple garden pool can form the centerpoint of your backyard or patio landscaping theme. Small, shallow pools like the one illustrated in this project require no permanent drain or supply lines. They can be filled and emptied in a short time by hand. Portable, submersible pumps can supply a steady stream of recirculating water to create gentle waterfalls or fountain spray jets. Follow these basic steps:

1. Dig a saucer-shaped excavation for the pool that is 6″ deeper at the center than the desired depth of the pond. If you wish to stock the pond with fish or plants, check with your nursery to find out how deep the pond should be. A foot deep is usually enough for fish; 16″ to 18″ is usually needed for plants. Make

Plywood   1 x 2 Border

Sand Bed

Methods of creating recessed or embossed lettering or shapes

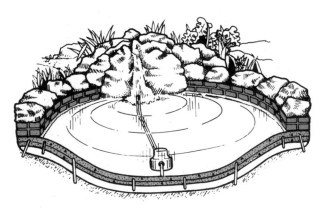

Shallow garden pools do not require water supply and drainage lines. A submersible pump recirculates water to create waterfalls and fountain jets.

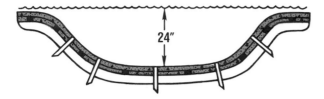

Grade stakes help position the layers of sand base and concrete and maintain the gently rounded slope of the pool.

sure that the depth at the edge of the saucer is 6″ below the ground surface, too. The sides of the pool should slope no more steeply than 45°, and preferably 30° or less, in order to prevent ice damage in the winter.

2. Cut a series of 11″ long stakes from 1×2 lumber, and mark each at the 2″, 5″, and 8″ distances from the head. Drive them into the bed of the pool at square foot intervals up to the 8″ mark.

3. Fill the pool area with 3″ of coarse sand or gravel as a subgrade (up to the 5″ mark on the stakes).

4. Lay wire mesh into the pool. Support it about 2″ above the subgrade with small stones or brick bits.

5. Pack the concrete firmly around the mesh up to the top of the stakes. Remove the stakes as you go. The pool must be cast all at once, so be sure you have enough concrete mix prepared. Mix the concrete with slightly less water

than in other construction so that it will not sag when placed on the sand.

6. After the concrete loses its sheen, finish it with the trowel or wooden float.

7. Cure the concrete for at least three to four days.

8. The perimeter of the pool can be built up using mortared brick or dry-laid or mortared stone. Plants and shrubs can also be used to create a cool, shaded effect. Painting or staining the surface of the pool bottom a dark navy blue or green will make the pool appear deeper.

9. If you plan on installing a submersible, circulating pump to create a tiered waterfall or spray fountain effect, follow the manufacturer's directions carefully. Keep in mind that running water does produce noise, so you may wish to limit the drop of the water or its volume. You should know what type of pump setup you are going to use before building the pool's stone or brick perimeter. Tubing or electrical cables will likely have to pass through or be concealed by these items.

10. Thoroughly clean and flush the surface of the pool before adding any plants or fish.

Float finish the pool bottom. Build up the sides of the pool using brick, stone, or concrete. Pitch waterfall ledges forward so the water flows back into the pool.

SETTING POSTS IN CONCRETE

Stable wooden and metal posts set in concrete form the backbone of dozens of successful, useful, and enjoyable home projects. You can mix post-setting concrete from scratch, but this job is ideal for more convenient premixes. In fact, special fast-setting concrete premixes are available that make setting posts as easy as possible. These fast-setting concretes require no mixing; water is added to the dry mix right in the posthole. Initial set is in fifteen minutes or so, and loads can be applied in several hours. With conventional concrete and concrete premix, loads should not be applied for one to two days.

Common uses for permanently set posts

Forming Postholes

Dig the posthole with an auger or a clamshell digger, rather than a shovel. A shovel makes the hole too wide. The diameter of the posthole should be roughly three times the post diameter. The hole depth should be a third of the overall post length. When posts are used for structural support (decks, etc.), the hole must extend sev-

eral inches below the frost line. Undercut the hole base to increase support strength.

Placing the Concrete

If a fast-setting premix is used, follow the manufacturer's specific instructions. In a typical installation, the dry concrete mix is poured into

If you mix your own concrete or use a standard premix, place the fresh concrete in the hole with a shovel. Tamp the concrete to prevent voids, and always use enough concrete so you can form an above-grade slope away from the post as shown. This directs rainwater away from the post.

FENCES

Perhaps the most popular home project involving the setting of posts is fence building. All fences, from simple two-rail designs to more elaborate split-rail and picket designs, need solid support.

Before beginning the work, check with local authorities to see if there are any ordinances, regulations, or zoning laws concerning the height, location, and materials for fences in your locality. Your property deed may also contain restrictions concerning the construction of fences. Also, be absolutely certain of your property lines; have the area surveyed if necessary. If you mistakenly build a fence on a neighboring property, the fence is theirs, and they have the right to remove it, paint it, etc.

Plot the line of the fence to avoid major visible obstacles, such as trees and boulders, and also make sure you are away from all underground utilities before digging postholes. In most cases, a call to your phone, gas, cable, and/

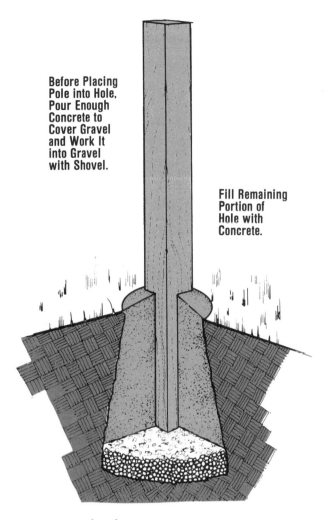

Before Placing Pole into Hole, Pour Enough Concrete to Cover Gravel and Work It into Gravel with Shovel.

Fill Remaining Portion of Hole with Concrete.

Post-setting details

the hole until it is half full. One-half of the recommended water content is then poured into the hole and allowed to soak into the concrete. The hole is then filled to the top with mix and the remainder of the water is added.

With fast-setting concrete mixes, water is added to the dry mix placed in the hole.

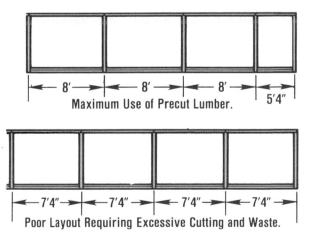

Maximum Use of Precut Lumber.
8'   8'   8'   5'4"

Poor Layout Requiring Excessive Cutting and Waste.
7'4"   7'4"   7'4"   7'4"

Examples of good and poor fence layouts

or electric company will prompt a free visit to your property to survey the situation.

Measure the fence's overall length, allowing space for one or more gates, if desired. Divide this length into equal intervals of 6', 8', or 10'. In this way, standard precut lumber can be used for the crossrails, minimizing cutting waste. If one section is smaller, consider using it as a gate location.

To find the exact post locations, mark both end points of the fence with wooden stakes and run a line between them. Locate positions for all posts between the end posts by measuring intervals with a tape, or by laying out the precut rails in line along the ground.

Usually 4×4s are a good size for fence posts, but 6×6s are required for unsupported corners and posts for heavy gates.

Dig the postholes in the correct locations and set the first end post in concrete in the manner explained earlier. Then, set the second end post firmly on its gravel base, but do not immediately pour concrete in its hole. Run a string between the tops of these two posts, and then position the interval posts. Carefully align and brace them vertically. Make sure the posts are correctly spaced and that the tops of all posts are level with one another.

Wrong

Right

Fencing Slightly Uneven Terrain.

Fencing Severe Slopes.

Handling slight and severe slopes

Mix and pour collars of concrete mix for all of the remaining posts. Double check for plumbness. Allow the concrete to cure three to four days before adding the rails and facing.

To finish the fence, attach the top rails or stringers first. These rails are usually placed flat on top of the posts to keep the fence in alignment. Various types of butt, lap, dado, and mortise-and-tenon joints can be used to fasten the rails to the posts. With the top rail in place, measure down the post to position the bottom and middle rails. Before attaching the rails to the posts, apply paint or wood preservative to the cut ends of the lumber.

Once the rails are in place you can add the facing. Facing can be done with wood patterns, pickets, wire screen, or solid panels.

If the ground is not level, make sure that you account for differences so that the fence is even. The ground should be built up and cut back so as not to give a choppy appearance.

On severely sloping ground, be sure to erect posts plumb to one another and then tilt the rails as needed to follow the slope. Tilting the rails changes the lengths required to span posts; so make sure you don't come up short when using standard lengths of lumber.

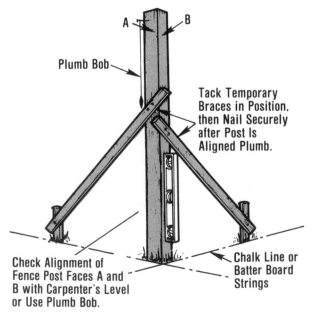

A B

Plumb Bob

Tack Temporary Braces in Position, then Nail Securely after Post Is Aligned Plumb.

Check Alignment of Fence Post Faces A and B with Carpenter's Level or Use Plumb Bob.

Chalk Line or Batter Board Strings

Checking post alignment

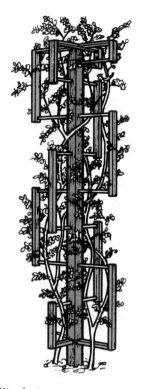

Common trellis designs

TRELLISES

A trellis to support perennial vines or roses can make a handsome addition to any garden, patio, or landscaped area. Trellises can stand alone, against a building or entranceway, or as part of a fence/windbreak construction.

Although young vines start out quite slender, lightweight, and fragile, by the end of a full growing season they can become extremely heavy and full. Use posts of the dimensions used in fence construction, and attach sturdy rail and crosspieces.

Be sure to use a nontoxic wood preservative to treat the trellis. Do not use creosote. Apply three coats of quality outdoor paint. Consider matching the color of your house or that of the vine blossoms.

Chapter 9

STUCCO AND SPECIAL CEMENT-BASED PRODUCTS

Stucco is a cement-based coating with many of the desirable properties of concrete. When applied correctly, it is hard, strong, durable, and weather resistant. It is unaffected by rust and insects. Thus, maintenance is inexpensive. In addition, a multitude of decorative finishes is possible with stucco because it can be colored by adding mineral pigments during the mixing process, and its surface can be given an unlimited variety of texture and patterns.

Stucco is usually applied in three coats: a scratch coat, brown coat, and finish coat. Both the scratch coat and brown coat are made of a mortar mix similar to that used in brickwork and blockwork. Prepackaged mortar mixes can be used, or you can mix from scratch using one part portland cement, three parts masonry sand, and one-quarter part hydrated lime. Lime is needed to give the mix the stickiness and cling necessary to adhere to vertical surfaces.

The finish coat can be mixed from scratch using white silica sand and white portland cement and lime in the above proportions. However, in the vast majority of stucco work, a prepackaged stucco finish coat mix is used. Most stucco finish coats are white or light gray, but colored finish coats are also available. White stucco can also be pigmented, painted, or stained with products formulated for use on cement-based surfaces.

Although stucco is used primarily for exterior walls, it is also well suited for interior locations such as kitchens, laundries, saunas, bath-rooms, or anywhere there is a high degree of moisture in the air.

APPLYING STUCCO

Stucco can be applied to concrete, masonry, or wooden surfaces. Good workmanship and attention to detail is important. Using shortcuts or slipshod techniques is liable to result in a cracked finished surface.

Apply stucco in mild weather (50° to 80° F), and work when the surface is shielded from the hot, baking sun that could dry the stucco too quickly and cause cracking. Hang shade cloths from the eaves if necessary. If you are doing stucco work as part of a new home construction, hold off until all interior wallboard is nailed in position. Large amounts of hammering could loosen a freshly stuccoed surface from the lath.

Preparing the Surface

Concrete or cinder block masonry provides an excellent base for stucco. These walls are very rigid, and their slightly open texture promotes strong bonding. Placed concrete walls are also a good base for stucco, provided the surface is not overly smooth. You can test the surface of concrete, masonry, or old stucco by spraying it with clean water. Note how quickly the moisture is absorbed. Porous, strong-bonding surfaces will absorb water quickly. But if water droplets form and run down the surface, the wall may be too smooth to accept the stucco coating. A coating of concrete bonding adhesive brushed or sprayed onto the surface and allowed to dry overnight may sufficiently increase the bond on slightly rough surfaces. Extremely smooth concrete can be roughened by sandblasting. You can also treat smooth concrete like wood, attaching roofing felt and metal lath as a backing for the stucco.

Concrete and masonry surfaces should be free of oil, dirt, or other materials that could weaken the bond. Painted concrete or masonry must be sandblasted to remove the paint and improve the bonding characteristics.

As shown, wooden walls use a base of waterproof roofing felt that is covered with fine screen metal lath. The roofing felt is fastened to the sheathing with galvanized nails. Each horizontal strip of felt should overlap the one below it by 3″.

Metal stop beads are installed along the base and the top of the wall and around window and door openings. These beads support the lath and stucco and allow moisture to escape from behind the stucco. Stop beads are formed in 10′ lengths that can be cut to size using metal snips or a hacksaw. The beads are nailed to the concrete foundation or the edge of the wood sheathing. Be sure that the beads are installed level and that separate sections of bead are aligned with one another. Special corner beads are installed

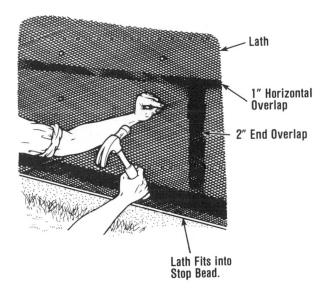

Applying wire lath

at outside corners to create true, even corners at these points.

When attaching the lath, it is best to work from the bottom up. The first lath sheet should rest in the stop bead. Every successive sheet should overlap 1″ of the sheet below it along the horizontal line. The vertical ends of the sheets should overlap 2″. The metal lath should extend 3/4″ beyond the corner beads at the outside corners.

The metal lath does not rest directly on the roofing felt. It is attached using special self-furring nails that create a 1/4″ space between the lath and felt. This space is absolutely essential for good adhesion. Self-furring nails are designed with a 1/4″ long plug surrounding the nail shaft. The plug is positioned behind the lath and against the wall. The lath is hooked over the nail head and the nail is driven into the wall.

It is easiest to work from the center of a metal lath strip out to the edges. Space the furring nails every 6″ in all directions. Have a helper pull on the lath to keep it taut during nailing. This prevents any bulges or depressions in the lath. Some types of lath are extremely sharp, so wear protective gloves, if needed.

Self-furring lath is also available. This product combines the lath and building felt backing

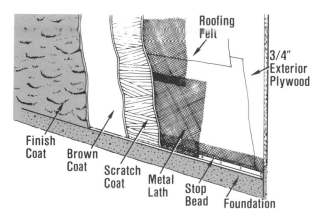

Details of a three-coat stucco system applied over exterior plywood

in a single roll. It installs using ordinary roofing nails.

As the final step of preparation you should apply masking dope to door and window castings to protect them as stucco is applied.

Applying the Scratch Coat

Using enough water to achieve a putty-like consistency, prepare as much mortar as can be used in about an hour. Next, trowel the mortar onto the surface from bottom to top on concrete or masonry surfaces, and from top to bottom on wire mesh. Make certain that concrete and masonry surfaces are damp before applying the mortar. The scratch coat should be about 1/2" thick. Force it through the wire mesh on wooden structures with an upward motion so that it is about 1/4" thick behind the mesh. Smooth the scratch coat with the trowel. Do one complete wall at a time. Do not try to get tight joints around windows and doors. The caulk will take care of sealing the joints.

Check carefully for any pockets or bulges in the coating and correct them before the stucco dries. After the coat has set enough to be firm but not hard, use a rake to scratch horizontal grooves about 1/8" deep across the face of the mortar. Do not rake the scratch coat so hard that stucco is removed. Only rake hard enough to score the coat. Allow the scratch coat to dry

Grooving the scratch coat

overnight. With open-frame wood construction, do not apply the brown coat for at least forty-eight hours after applying the scratch coat. As with all concrete, slow, damp curing provides the most strength, so try to keep the stucco damp by misting with a garden hose periodically during that time.

For best results, do not apply stucco to a frozen surface or when freezing temperatures are likely.

Applying the Brown Coat

The second coat is called the brown coat because it used to contain brown sand. Special care must be taken to make this coat smooth and even because the final coat is only 1/8" thick and it will not hide any irregularities. You can ensure that the surface is even by stretching strings horizontally across the wall and attaching them to nails placed beyond the corners. The strings should be held out 3/8" from the scratch coat by the corner beads. Place one string at the top, another at the bottom, and one or more in the middle section. Roofing nails should be driven in every 5' along each string so the heads are flush with the inside of the string. Once the nails are in place the strings can be removed. The nails serve as guides to ensure that the coat is level.

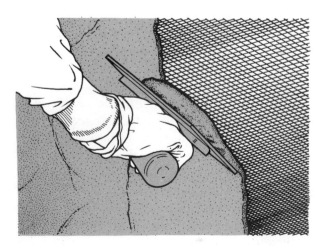
Applying the scratch coat

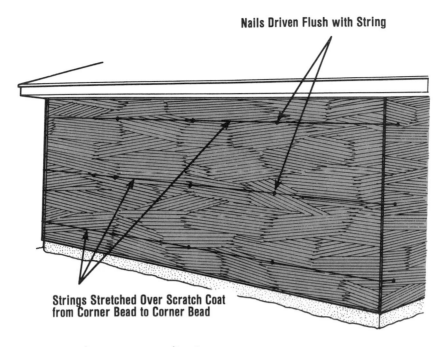

Nails Driven Flush with String

**Strings Stretched Over Scratch Coat
from Corner Bead to Corner Bead**

String lines ensure an even brown coat application.

A

B

Applying the brown coat: (A) troweling on the mortar and (B) screeding it level

Dampen the surface of the scratch coat evenly with a fine spray of water, but do not soak it. Prepare as much mortar mix as can be used in an hour. Use a float and hawk to apply the brown coat to a thickness of 3/8", just thick enough to cover the nail heads. Do the entire wall at one time; otherwise, a color difference might be seen in the finish coat where you stopped and started again. Screed the brown coat evenly by running the edge of a straight board over the coat and smoothing out the high and low spots. After it loses its sheen, float it smooth. Wait about half a day and then moist-cure the brown coat by dampening it with a fine spray of water every few hours for two days. Then air-cure the surface for five more days.

Applying the Finish Coat

Trowel on a 1/8"-thick coat of prepared stucco finish, starting at the bottom and working to the top. Do an entire wall at one time for a consistent finish.

You can texture the finish coat to suit your own tastes. For a smooth, plasterlike finish, trowel the final coat several times as it becomes progressively stiffer. A swirling texture can be produced by troweling the stucco just once with an arcing motion and allowing the resulting

pattern to remain. You can also use the bristles of a hand broom to create accents and wavy patterns in smooth troweled stucco after it has hardened slightly. Another light finishing technique is to float the finish coat, allow it to stiffen slightly, and then scrape the finish very lightly with a short length of 2×4 lumber. Use vertical strokes and press lightly.

Jabbing the surface with the bristles of a broom will create a rough, raised texture. After the finish coat has set up slightly, these ridges can be flattened with a steel trowel to create a travertine finish. Small globs of finish coat can also be applied with a round nose trowel and pressed semiflat to create an interesting accent pattern.

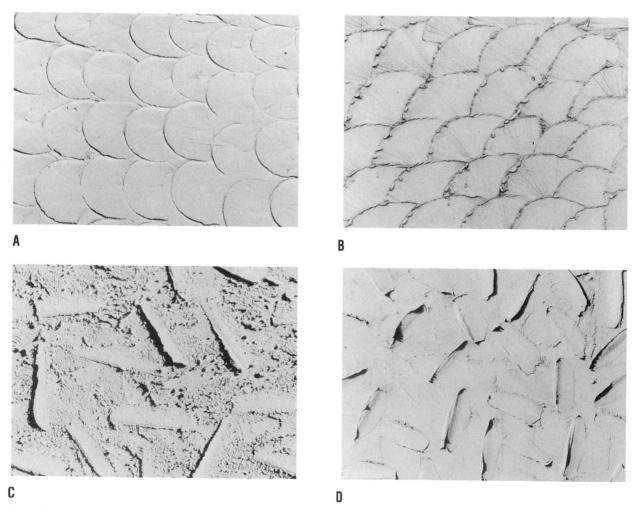

A   B

C   D

Sample stucco finishes: (A) steel-troweled swirl, (B) steel-troweled shell with hand-broomed highlights, (C) travertine finish with round nose pointing trowel highlights, and (D) rough floated finish with flattened highlights.

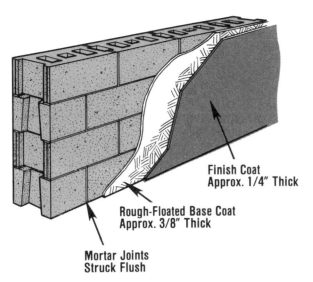

Finish Coat
Approx. 1/4" Thick

Rough-Floated Base Coat
Approx. 3/8" Thick

Mortar Joints
Struck Flush

Two-coat stucco applied directly to concrete masonry

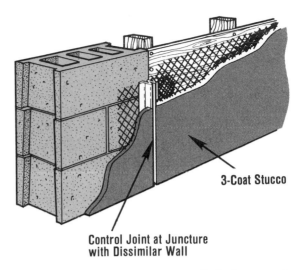

3-Coat Stucco

Control Joint at Juncture
with Dissimilar Wall

Stucco control joints must be installed wherever there is a control joint in the underlying wall or at points where walls of dissimilar materials meet.

You can also impress designs into the surface. Slap a leafy branch against the surface or impress some other pattern into the soft surface and then trowel the surface lightly.

Moist-cure the finish coat for five days by misting it periodically with a fine spray. Then seal all the joints around doors and windows with a masonry-compatible caulk.

Two-Coat Applications

While stucco is usually applied in three coats, there are instances where two coats will suffice. Horizontal or overhead applications seldom exceed two coats, and two coats are often used when stucco is applied directly to concrete masonry, which provides a very rigid and stable base. In these cases, the brown coat is eliminated and the finish coat is applied directly to the scratch coat. The scratch coat should be completely dry before the finish coat is applied.

Installing Control Joints

Undesirable cracking can be substantially minimized by careful preparation and application of the stucco. Nevertheless, cracks can develop through many causes or combinations of causes, which include:

- Shrinkage stress
- Building movements
- Settling foundations
- Construction joints
- Intersecting walls or ceilings, corners, and pilasters
- Restraints from lighting and plumbing fixtures
- Weak sections due to cross-section changes such as at openings

While it is difficult to anticipate and prevent cracks from all these possible causes, they can be largely controlled with the help of metal control joints. These control joints should be installed directly over any previous joints in the base.

Walls and ceilings that use metal lath to anchor the stucco should be divided into rectangular panels with a control joint at least every 20'. The metal lath should not extend across these control joints. Use metal that is weathertight and corrosion-resistant for control joints on exterior surfaces.

Coloring Stucco

As mentioned earlier, if left untinted stucco finish coat cures to a white or light gray finish.

Cleaning cracks in stucco with a small knife

Brush away all dust.

Prepigmented finish coat mixes are also available in brown, beige, and pastel colors. You can also tint white or light gray finish coats during the mixing process by adding mineral-oxide pigments. Only oxide pigment colors are stable over a long period of time. As in coloring concrete, carefully measure and record the amount of pigment added to each batch. Keep proportions consistent throughout the entire job. Mix thoroughly, using clean tools and mixing vessels.

Painting or staining the finish coat is another option to consider. Painting normally requires a priming coat compatible with masonry surfaces followed by a top coat of latex-based paint. Pigmented stains that penetrate into the surface of the finish coat can also be used. It is extremely important that the stucco is adequately cured before the paint or stain is applied. The waiting period for some paints and stains can be as long as sixty days, so follow all manufacturer's guidelines to the letter to guarantee success.

REPAIRING DAMAGED STUCCO

One of stucco's advantages as siding is that it rarely needs repairs, and when it does need repairs they are usually simple. Repair procedures for cracks and holes in stucco are given here.

Repairing Cracks

Follow these steps to repair any cracks in the stucco.

Apply the patch and finish to match the surrounding texture.

1. Remove any loose stucco from the damaged area.
2. Use a hammer and chisel or knife to undercut the stucco surrounding the damaged area so that the patch will be locked in.
3. Use a wire brush to remove any stucco knocked loose in undercutting. Brush the entire damaged area free of dirt, dust, and other foreign materials.
4. Thoroughly wet down the damaged area and keep it damp for twelve hours before patching to prevent the moisture from being drawn too quickly from the patch and weakening it.
5. Mix the stucco finish mix with an acrylic fortifier. Tightly pack the prepared stucco finish into the damaged area with a trowel and texture it to match the surrounding stucco.

6. For deep repairs extending to the scratch coat, first tightly pack the hole within 1/4" of the surface. Keep the patch damp for two days while it sets.
7. Dampen the edges of the hole and the surface of the first layer; trowel in a second layer level with the original stucco and texture it to match.

Repairing Holes

To repair large holes, take the following steps:

1. Remove the stucco all the way down to the wall surface in the area to be repaired.
2. Weave in new wire lath.
3. Dampen the area and apply the scratch coat. Score the coat once it begins to set, then keep it damp for the next forty-eight hours.
4. Apply the second coat so that it comes within 1/8" of the surface. Keep it damp for two days, then air-cure it for five more days.
5. Dampen the wall and apply the finish coat. Remember to add color pigment if any was used in the original surface. Keep the surface damp for two days.

Adding a liquid acrylic fortifier as part of the concrete mix

CONCRETE AND MASONRY SEALERS

Clear acrylic concrete and masonry sealers are often applied to protect both interior and exterior concrete and masonry from damaging acids, alkalis, household cleaners, gasoline, oils, and salts. The acrylic-based sealers form a clear, extremely weather-resistant finish that is ideal for driveways, patios, and garage and basement floors. They seal surfaces against dusting, and retard spalling and weathering. Some treatments are also recommended as a masonry primer for latex paints.

ACRYLIC FORTIFIERS

Water-resistant acrylic resins are available to increase the overall strength of concrete mixes. These liquid admixtures are added as part of the water portion of the mix. They are also commonly used to strengthen stucco and other surface-bonding cements.

Acrylic fortifiers may also be used to bond new toppings up to 1" thick. When the topping

Clear acrylic sealers can be applied to protect all types of concrete and masonry surfaces.

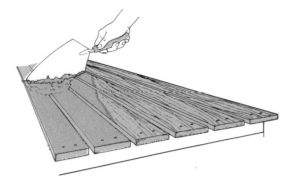

Troweling on the skidproofing mix

To Texture, Move the
Brush or Broom Sideways,
Going in the Same
Direction with Each
Stroke.

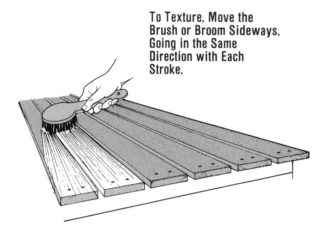

Creating a brushed texture

exceeds 1″ in thickness, it is best to use a surface-bonding adhesive as described in the next section. This is because acrylic fortifiers are air-hardening products. When the overlay is too thick, the acrylic will not harden at the bottom of the overlay, the bond will not take, and the concrete will not reach its full strength. The applications outlined here are typical. Always follow the manufacturer's specific instructions for use and application.

Skidproofing Wood, Metal, and Concrete

Acrylic fortifiers can also be added to sand mix (three parts to one part portland cement) to create a nonskid surface around pools, boat docks, and so on.

A brushable concrete-bonding adhesive should be used to ensure a strong bonding patch or new top coat.

1. Lay masking tape around the edges of the surface to be coated. Make certain that the surface is clean and free of dust, dirt, and loose matter.
2. Mix 1-2/3 quarts of fortifier with 60 pounds of sand mix; for larger and smaller mixtures refer to the directions on the back of the fortifier bottle. Add water to achieve the desired consistency. Acrylic paint can be added to the wet mix before application if a particular color is desired.
3. Trowel the mix onto the surface in a 1/16″- to 1/4″-thick coat.
4. Brush the mix to achieve a rough texture by moving the brush sideways, going in the same direction with each stroke.

BONDING ADHESIVE

Concrete bonding adhesive penetrates pores of old concrete and forms a chemical bond for new concrete toppings up to 2″ thick. Use the bonding adhesive rather than an acrylic fortifier when the thickness of the topping is above 1″. This adhesive provides a strong chemical bond for applications of plaster, stucco, gypsum plasters, and similar materials. Bonding adhesive is good for some outdoor uses, but not for all. If a concrete overlay is being placed on concrete, the adhesive can be used. Water won't penetrate through the concrete and the adhesive will be kept dry. In areas where water penetration is likely, such as stucco patcher 1/2" or 1/4" thick, a bonding adhesive may have problems holding the patch. However, the adhesive will hold up well if a sizable overlay is being placed over concrete, even outdoors.

To use the adhesive, just brush it on and wait until the surface becomes tacky or dries. Roughing the old surface is usually not necessary. When the adhesive is dry, it will no longer be a white film in appearance; instead it will be clear. Then, you can put on the overlay or wait a day to a week. It doesn't really matter how long you wait to apply the overlay, so long as the adhesive coat doesn't get dirty or rained on. But it is better not to wait more than two weeks to put on the patcher. That is about the extreme outside limit. Most bonding adhesives clean up with soap and water.

HYDRAULIC CEMENTS

Hydraulic cement is a special mixture of portland cement, calcium aluminate, fine sand, and special additives. It sets quickly, and actually stops flowing water in three to five minutes with its characteristic expansion and high early strength. Because it sets so quickly, hydraulic cement cannot be used for patching that needs to be sculpted. Any patching that needs to be done had better be done quickly. Hydraulic cement sets so quickly that the powder doesn't have a chance to dissipate underwater as other portland products do. The cement expands very slightly in order to lock into the opening and form a tight seal against any further leaking. Hydraulic cement is suitable for plugging leaks in masonry walls and/or patching concrete basements and retaining walls, cisterns, swimming pools, fountains, etc. This material is also excel-

Applying a hydraulic water-stopping cement to the base of a leaking basement wall (Courtesy of Harry Wicks Photos)

Wrong   Good   Best

A properly shaped and prepared hole allows the expanding hydraulic cement to grip tightly

lent for caulking; tuckpointing; repairing masonry chimneys, walls, and sills; and for sealing openings around pipes and fixtures.

When applying the hydraulic cement, all patch areas should be free of loose material, dirt, and dust. Preparation should include enlarging all cracks and holes, and avoiding "V"-shaped cuts. Starting at the top and working down, apply the cement while maintaining light pressure on the patch. Maintain pressure until the initial set begins and the leaking stops.

Mix the hydraulic cement according to manufacturer's directions. Normally, ingredients are

When plugging deeper holes and cracks, form the cement into a plug and push it into place.

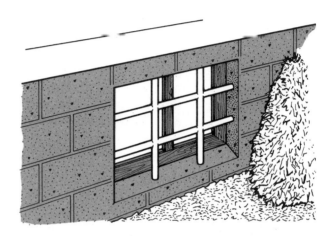

Anchoring cements can be used to securely mount hooks, fixtures, and other items in existing concrete and masonry.

mixed to form a heavy putty consistency. The rapid set time requires the user to limit the batch size during application. Mix up only as much as can be used in a few minutes. If you are working on a leaking hole, mix smaller portions because you will be working more slowly; one hand will probably be needed to hold the plug in place while setting. When a faster set is desired, a warmer mixing water should be used. Higher water temperatures produce faster sets.

ANCHORING CEMENTS

Anchoring cement expands as it hardens to become stronger than concrete. Use anchoring cements for setting bolts, posts, hand rails, machinery, fences, columns, and almost anything else to be anchored in concrete.

When using an anchoring cement, place it in cuts with vertical sides. It is a rapidly expansive material and would just pop out of a hollow depression. After adding water, it pours like syrup and normally sets in ten to thirty minutes. In an hour, it will be hard enough to screw any type of bolt home. For extremely high tension, you may want to wait two to three hours. This cement can also be made to a packable, mortar-

like consistency for vertical patching, such as in walls, by just adding less water. Use only clean water and containers to prepare the mix.

One limitation with anchoring cements is their potential to be adversely affected by constant water exposure. Normally, however, this will not be a problem when the anchoring cement is inserted in concrete. The concrete will protect it, keeping it dry all around except for the surface. Just be sure to coat the surface with a waterproofing concrete and masonry sealant.

Floor Mountings

To use anchoring cements for floor mountings, follow these steps:

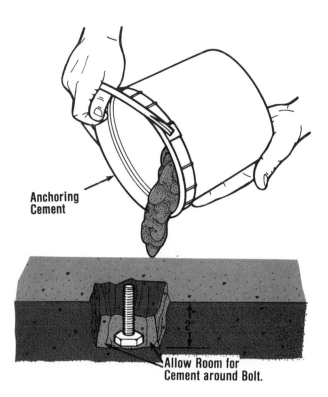

Anchoring Cement

Allow Room for Cement around Bolt.

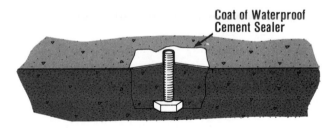

Coat of Waterproof Cement Sealer

Details for typical floor mounting using anchoring cement

1. Mark the location of the holes for the bolts or posts.
2. Make a hole at least 1″ larger than the diameter of the object to be anchored. For large objects such as metal pipes or fence posts, use a diamond-edge cold chisel and a small sledgehammer to make the hole. The hole must be at least 2″ deep.
3. Brush all dirt, dust, and other loose material from the hole. Dampen the interior of the hole. Apply concrete bonding adhesive to holes with smooth inte-

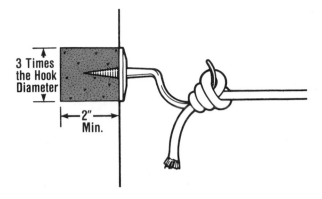

3 Times the Hook Diameter

2″ Min.

Details for typical wall mounting using anchoring cement

rior surfaces before applying the anchoring cement.

4. Insert the object to be anchored and pour the prepared mix into the hole to the surface level. Anchoring cement can also be packed in with a trowel; when using this method, be sure to completely fill the hole.
5. Hold the object being anchored in place until the mix begins to stiffen. This will depend on the amount of water used to prepare the mix.
6. Wait about forty-five minutes before fastening anything to the bolt or post to allow the mix time to achieve a final set.

Wall Mountings

To mount objects in a wall, such as brackets, hooks, racks, shelves, pulleys, and awning and canopy mounts, follow these steps:

1. Mark on the concrete the location of the hole(s) for the hook or mounting bolts.
2. Make a hole at least 2″ deep and about three times the diameter of the bolt or hook to be inserted.
3. Brush all dirt, dust, and other loose material out of the hole.
4. Dampen the interior of the hole; leave no standing water. Insert the hook or

bolt and pack the prepared mix firmly around it to just above surface level. Prepare only as much mix as can be applied in ten minutes.

5. Hold the bolt or hook in place until the mix begins to stiffen. (It generally takes only a few minutes. Mix prepared with large amounts of water will take longer.)

6. After all the bolts have set for about forty-five minutes, attach any mounting brackets that your project requires. Except for the most unusually heavy objects, mounting can be done after a few hours. Anchoring cement has a typical compressive strength of 3,000 psi after two hours.

QUICK-SETTING CEMENTS

These specialty cements are formulated to set hard in only five to ten minutes. They are used to make repairs where high strength and rapid setting are needed such as on damaged swimming pools, septic tanks, retaining walls, precast concrete pipe, well covers, culverts, curbing, steps, and other concrete units such as birdbaths and sundials. The main advantage of

this type of product is that it can be shaped, shaved, and sculpted during application. Formwork is not needed. See page 158 for additional information.

MASONRY COATINGS

The term *masonry coatings* is ordinarily used in reference to products used to moisture-proof, seal, and color porous masonry surfaces, indoors and outdoors. Unlike ordinary paints or stains, masonry coatings contain certain amounts of portland cement or other water proofing agents. They can be purchased as a liquid or white or tinted powder that is mixed with water to produce a thick paint-like liquid that can be applied to poured concrete, concrete block, stucco, brick, or stone. Masonry coatings are usually acceptable for above and below grade applications. They are less expensive than paint products and contain no combustible materials.

In addition to providing dampproofing, masonry coatings are excellent for filling hairline cracks and restoring good surface appearance. Application thickness is approximately 1/16". The use of acrylic fortifier in the mixing water will increase coating strength and bonding ad-

Quick-setting cements can be shaped and sculpted for edge and ornamental repairs.

Masonry coatings and waterproofers are available in liquid and powdered mix form. They are usually brushed on like heavy paint.

Surface-bonding cements are applied similar to a stucco finish coat for mortarless block construction and renovation.

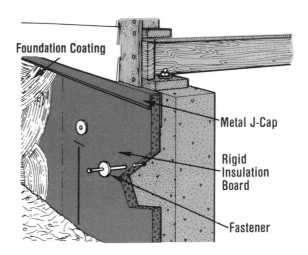

Details for insulating with rigid insulation board and special foundation coatings

hesion. Prior to application, clean the surface of all loose material, dirt, dust, flaking paint, grease, and so on. Refer to chapter 10 for complete surface cleaning and preparation instructions.

After cleaning, any smooth surfaces should be roughened or etched, and the entire area should then be dampened with water. When applying during hot weather, the walls should be lightly misted several times during the day to eliminate rapid moisture loss. Water quantity per unit must be consistent to ensure color uniformity. A two-coat application can be used to achieve a greater coating thickness. Apply the second coat twenty-four hours after the first. Do not moisten the wall after the second coat. When mixed to the proper consistency, the masonry coating mixture may be applied by spraying or with a brush or paint roller.

SURFACE BONDING CEMENTS

These cement-based blends of fiberglass and fine sands can be used to bond together dry-stacked concrete block for simple, mortarless construction (see chapter 15). Surface bonding cements can also be used in repair and renovation work as a water-resistant seal and decorative finish over existing concrete or masonry

walls (see chapter 10). Special colored blends are available.

FIBERGLASS-REINFORCED FOUNDATION COATINGS

Another premixed specialty product, this fiber-reinforced cement-based material provides a protective exterior coating over below-grade exterior rigid board insulation.

So if your home has uninsulated, poured concrete or concrete block basement walls, you can reduce heat loss substantially by installing rigid insulation and coating it with this brushable product.

Insulation panels can be extruded polystyrene foam (blue board), expanded polystyrene (white bead board), urethane panels, pressed glass fiberboard, or other rigid insulation materials. It is best to insulate to the frost line, but substantial savings are possible even if insulation is applied from the bottom of the siding to a depth of 1' below grade.

Wall Preparation

Determine the depth to which you are going to insulate. One foot below grade should be the

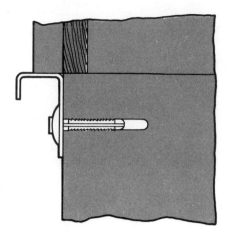

Attaching the metal J-channel along the top edge

minimum. To estimate the amount of rigid board needed, measure from the bottom edge of the siding to the below-grade depth and multiply by the perimeter of the house. This will give you the square footage of board needed. To prepare for installation do the following:

1. Dig out the soil from the foundation to the required depth.
2. Apply a bead of caulking to the underside of the bottom edge of the siding.
3. Nail or fasten the metal J-channel to the foundation, pressing the channel into the caulk to prevent water from flowing behind the channel and insulation. Fasten the channel beneath the edge of all siding and around all door and window openings. (See chapter 18 for details on fastener selection.)

Installing the Insulation

The rigid board can be secured to the foundation with special concrete fasteners and washers or with panel and foam adhesives specially formulated for use with rigid insulation board.

1. Measure and cut the insulation panel to size using a saw or utility knife. (With certain types of extruded polystyrene boards, the clear plastic coating must be

removed prior to installation.) Make certain all panels butt tightly against one another and the J-channel.
2. If securing with fasteners and washers, predrill the proper size holes through the rigid board into the foundation on 24" centers both vertically and horizontally. Holes must be deep enough to accommodate the fasteners and allow the washers to be recessed into the panel. Butt the panels at the corners so one panel overlaps the other.
3. If securing with an adhesive, use only those adhesives designed to be used with rigid foam insulation. Solvent-based construction adhesives will destroy the installation board. Apply the adhesive according to the manufacturer's instructions.
4. Apply self-sticking fiberglass tape to the joints between the insulation panels, to all outside corners, and over all fastener heads and damaged panel areas. Brush dirt and residue from the insulation surface.
5. Mask around channels, windows, and doors to prevent contact with the masonry coating.

Applying the Coating

After the insulation board is in place, follow these steps to apply the fiberglass-reinforced foundation coating.

1. Mix the fiberglass-reinforced foundation coating according to package directions. You might want to slightly increase the amount added to set a more "brushable" consistency. If the mix begins to stiffen during work, remix, and add small amounts of water as needed.
2. Use the trowel to precoat all taped areas with mix.
3. With a brush or trowel, cover the insulation panels to the desired thickness. To prevent color variation, cover the entire

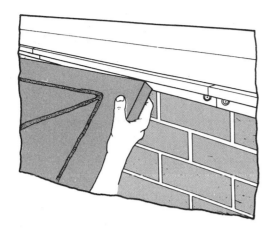

Install the rigid board with an adhesive designed for this application. Do not use solvent-based adhesives.

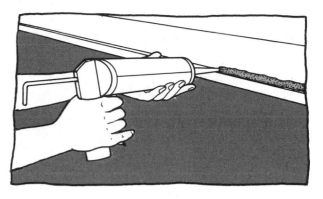

Caulk the top seam to ensure water-tightness.

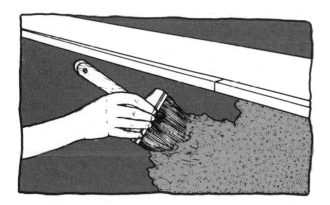

Brush the foundation coating onto the rigid insulation.

wall corner-to-corner in one work session. Clean the brush or trowel periodically to avoid a buildup of coating.
4. The final coating thickness should be approximately 1/8", although a thicker

coating can be applied if the surface must withstand abuse. The final coat is best applied with a stiff-bristled brush.

Finishing

To complete the project, follow these steps.

1. Remove all masking tape.
2. Replace the dirt from around the foundation, sloping it so that water runs away from the foundation.
3. Caulk around channels, doors, and windows to prevent moisture from penetrating behind the coating.
4. If desired, the coated panels can be painted with a waterproof masonry coating (see page 146).

Chapter 10

CONCRETE MAINTENANCE AND REPAIR

Although it is true that concrete is one of the most long-lasting and durable of all building materials, no material is 100% maintenance-free. Driveways and garage floors are stained with grease and oil. Paint and stains from general home upkeep and other projects find their way onto concrete and masonry surfaces. Mildew grows in damp, shady locations, and little can be done to prevent naturally occurring efflorescence.

Even the finest mixed and placed concrete is subject to natural forces that can crack and shift it. Accidents happen. A dropped heavy object chips a step. Careless power mowing does the same for a walkway or patio edge. Many times, first-time homeowners buy into the problems associated with old, poorly maintained concrete. The cost and effort of replacement often makes concrete repair and restoration the logical choice.

CLEANING AND STAIN REMOVAL

Concrete requires cleaning for two reasons. The first is to improve its appearance either by removing an isolated stain or performing a general cleanup over a wide area. The second reason concrete is cleaned is to prepare it for painting, top coating, patching, or any type of surface treatment or repair.

Methods of Cleaning

Stains and dirt can be removed from concrete and masonry by either dry (mechanical) or wet (chemicals and water) methods. It is sometimes necessary to use a combination of the two.

Common dry cleaning methods include scouring, grinding, sandblasting, and flame cleaning. Grinding, sandblasting, and flame cleaning require special tools, safety equipment (eye, breathing, and hearing protection), and procedures. Don't attempt these methods unless you are trained in the safe operation of this equipment.

Scouring should be done with nonmetallic brushes and pads to avoid rust discoloration.

Scouring involves some hard work, but no special skills. Steel-wire brushes and pads should be used with care. They can leave metal particles on the surface that will later rust and discolor the concrete. Always wear eye protection and heavy work gloves when scouring concrete.

Wet cleaning involves the application of specific cleaners or chemicals in diluted or concentrated solutions. Chemicals and cleaners work by either dissolving the stain or setting off a chemical reaction that changes the stain into another substance less noticeable or more easily removed.

Working Safely

CAUTION

Most chemicals used for cleaning concrete and masonry are hazardous and toxic. They require the use of specific safety procedures.

- *Avoid skin contact and breathing chemical fumes. Wear rubber or plastic gloves and chemical safety goggles. Work in well-ventilated areas, and avoid splashing or spraying chemicals, particularly when scouring.*
- *Follow storage and handling instructions given on the manufacturer's label. Unused portions that have been taken from the original container should be discarded in an environmentally safe manner. Never return them to the original container.*
- *Never store chemicals in open or unmarked containers. Keep all chemicals out of the reach of children, preferably under lock and key.*

Plan the cleaning procedure carefully. Remove or protect glass, wood, or metal items near the work site. Contact with some of the stain-removing chemicals could damage them. Always attempt to clearly identify the stain before selecting a cleaning agent. Use of the incorrect chemical or improper application of it can result in spreading the stains over larger areas. When identification is impossible, experiment with different solutions in a small area.

Keep in mind that removing stains from older concrete might leave that area much brighter than surrounding sections.

Although many chemicals can be applied directly to concrete without harming it, acids or chemicals that produce a strong acid reaction should be avoided. Even weak acids can etch and roughen a concrete surface if left on for an appreciable amount of time. Soak the area in water before applying an acid solution.

Chemicals for removing concrete stains can be purchased from sources such as commercial chemical supply houses, drugstores, hardware stores, and home center stores. Buy from a supplier who is knowledgeable in the field. Explain your needs and plans, and follow all safety advice and tips offered.

Mix acid solutions carefully and to the recommended strength. Always add small increments of acids to larger volumes of water; never pour water into a strong acid solution.

Cleaning Large Areas

The simplest method of removing dirt, grime, smoke, or general discoloration from concrete or masonry surfaces is to scrub it with a household scouring powder or detergent. You can create a stronger cleaning solution by adding 1/2 cup of trisodium phosphate (TSP) and 1/2 cup household detergent to 1 gallon of warm water. Be sure to wear rubber gloves and goggles. Soak the areas with water before scrubbing with the solution and rinse thoroughly after cleaning. Begin at the tops of walls and sloped slabs so that water and detergent run down over dirt below. The force generated by a hose spray nozzle might also be sufficient to remove loose dirt, paint, and chipped concrete.

Muriatic Acid Washes. A weak solution of muriatic acid consisting of one part muriatic

A

B

Removing stains with a chemically treated poultice: (A) adding solvent to the inert material, and (B) mixing the wet poultice.

(hydrochloric) acid mixed with twelve parts water can also be used to clean and brighten brickwork. This solution is particularly effective for removing fresh mortar smears and stains from new brick construction. Muriatic washes can also be used to brighten exposed aggregate surfaces. Muriatic acid is not normally used to clean concrete or concrete block surfaces because the acid etches or roughens the surface.

High-Pressure Water Cleaning. High-pressure water cleaning equipment can be rented to effectively clean large areas in a surprisingly short time. High-pressure washers generate a force of 5,000 to 10,000 psi, which is sufficient to blast away dirt, loose paint, and flaking concrete. They also effectively remove many stains, algae, and efflorescence.

Misuse of high-pressure cleaning systems can lead to serious injury and/or property damage. Never direct the spray toward yourself or others. The force of the water jet can enter your skin and cause serious complications. The forces created at high settings and pinpoint spray patterns can be destructive to concrete and other materials. Work cautiously at the lowest setting possible to do the job effectively. Wear goggles and protective clothing.

Sandblasting. This method uses compressed air to drive an abrasive grit onto the concrete or masonry surface. Although it effectively removes all dirt, paint, and debris, sandblasting also removes part of the concrete or brick surface. This may leave the surface, particularly older, soft brick, vulnerable to water penetration. On older brick homes, sandblasting may do more harm than good. It will also dull any shiny surfaces, such as exposed aggregate. However, sandblasting is an effective cleaning and preparation for concrete resurfacing and renovation because the rough surface it creates aids in bonding. Unless you have been trained in the use of sandblasting equipment and working procedures, hire a sandblasting contractor to do this type of work.

Chemical Cleaning. Large scale cleaning or paint removal projects using strong chemical cleaners should be done by a specialist. Materials used can be highly toxic and corrosive. Unless handled properly they can kill nearby shrubs, trees and grass, and damage wood and metal structures.

ISOLATED STAIN REMOVAL

Badly stained concrete and masonry require the use of more concentrated chemical solvents or reagents. You can apply the chemical

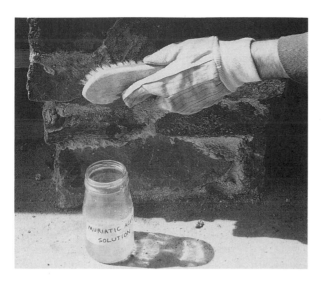

Cleaning stonework with a weak muriatic solution. Rinse the area thoroughly with fresh water after cleaning to remove all traces of acid that can damage some types of stone and brick over time.

by soaking a soft, absorbent cloth in the liquid and applying it as a bandage over the stain. As the chemical acts on the stain, the loosened particles are absorbed into the bandage.

A similar method involves the use of a poultice or paste made from the solvent or reagent and an inert, absorbent material. The paste or poultice is plastered over the stain and allowed to dry. As it dries, the active chemical in the paste dissolves the stain, which is then absorbed into the inert material. When dry, the paste or poultice is removed along with the stain.

The inert material can be talcum powder, whiting, hydrated lime, or fuller's earth. Selection of the solvent or reagent depends on the type of stain. One advantage of the paste method is that it prevents the stain from spreading during treatment. It also does a better job of pulling stains out of porous surfaces.

The following sections list removal procedures for many common stains.

Oil and Grease. Several manufacturers offer special concrete, masonry, and asphalt cleaners designed to remove oil, grease, and dirt buildup. Some, but not all, of these commercial cleaners offer the advantages of being nontoxic, nonflammable, and odor-free. Read all labeling and directions before use.

Removing grease stains with a commercial concrete cleaner

If an all-purpose cleaner is unavailable or unsuccessful in removing an oil stain, you can try one of several other methods. Try saturating the stain with mineral spirits or paint thinner, and then covering the area with talc, fuller's earth, and so on. Let it stand overnight, then sweep away the cover. Repeat this process if necessary.

You can also scrub the stain with a trisodium phosphate solution or bleach the surface with laundry bleach. Be sure to provide for good ventilation. If every other method fails, allow the concrete to dry, then apply a paste prepared from benzol and a dry, powdered material. Leave the paste on the stain for an hour after the benzol has completely evaporated, then remove it. This method can be repeated if necessary.

Blood Stains. Wet the blood-stained area with clear water and cover it with a thin, even layer of sodium peroxide powder. Take care not to breathe any of the peroxide dust nor to allow it to come in contact with the skin because it is very caustic. Eye and hand protection must be worn. Sprinkle the powder with water and allow it to stand for a few minutes. Wash with clear water and scrub vigorously. Next, apply vinegar to the surface to neutralize any alkaline traces left by the sodium peroxide. Finally, rinse with clear water.

Smoke and Wood Tar. Smoke and wood tar stains from barbecues, fireplaces, and wood burning stoves are difficult to remove. Begin with liquid laundry bleach or a solution of 1 cup ammonia and 1/2 cup water. Press the poultice firmly against the stained surface. Hold it in place with a concrete slab or sheet of plywood. Stains on vertical surfaces will require some creative support to hold the poultice in position. Commercial fireplace cleaning agents are also available.

Efflorescence. Efflorescence is a white, chalky film that appears on the surface of concrete and masonry. It is caused by certain soluble salts present in the concrete or brick that work their way to the surface. Most efflorescence salts are water soluble and can be removed by a high-pressure water jet, or dry brushing followed by flushing with clean water. Heavy accumulations or stubborn deposits can be scrubbed off with a diluted muriatic acid (one part acid to twelve parts water). Dampen the surface with clean water before applying the solution. After scrubbing thoroughly, flush with clean water to remove all traces of the acid. Commercial cleaners are also available to remove efflorescence.

Fruit. Fruit stains are organic and can be removed by scrubbing with powdered household detergent and warm water.

Graffiti. Many commercially available products are suitable for removing spray paint and felt-tip pen markings from concrete surfaces. These products are usually effective for removing crayon, chalk, and lipstick.

Algae and Lichens. These microscopic plants grow in damp, shaded areas. They can be killed by applying a household bleach solution or commercial herbicide.

Mildew. Commercial mildew cleaners are available, or you can mix your own cleaning agent using 3 ounces of trisodium phosphate crystals, 1 quart of household bleach, 1 ounce of laundry detergent, and 3 quarts of water. Scrub the affected areas with a soft brush and rinse with clear water.

Paint Stains. Different treatments are used for wet and dry paint stains.

For wet paint spills and spots, carefully soak up liquid paint with an absorbent material such as paper towels or soft cloth. (Avoid wiping since it will spread the stain.) Then scrub the spot with scouring powder and water. Keep scrubbing and washing as long as it makes an improvement. Then wait three days before resorting to a method for the removal of dried paint.

For old, hardened paint, scrape off as much as possible, and then apply a poultice impregnated with a commercial paint remover. After it has stood for twenty to thirty minutes, gently scrub the stain and wash off the paint film with water.

Paint that has penetrated the surface of concrete can be etched away using diluted muriatic or phosphoric acids. Heavier paint spots may require sandblasting or burning off with a propane torch.

Iron Rust. Remove rust spots and streaks by mixing 1 pound of oxalic acid in 1 gallon of water. Use a plastic or glass container, and wear eye and skin protection. Presoak the surface with water and scrub with a stiff fiber brush. Rinse the cleaned area thoroughly and repeat if needed.

Copper Stains. Runoff water from copper flashing or bronze fixtures can stain concrete a bluish-green or brown. To remove these stains, apply a paste consisting of one part ammonium chloride (sal ammoniac) and four parts talc, whiting, or clay. Aluminum chloride can also be used in place of the sal ammoniac. Allow the paste to dry before removal. Repeat as often as needed, finishing by scrubbing the area with clean water.

Aluminum Stains. Aluminum stains appear as a whitish deposit on the surface of the concrete or masonry. They can be scrubbed off using a diluted muriatic acid wash. Rinse thoroughly to avoid etching concrete surfaces. The dissolved aluminum salts may reappear as efflorescence. If this occurs, repeat the cleaning process.

Caulking Compounds. Scrape off as much of the caulking as possible and then apply a poultice of denatured alcohol and inert material. This treatment dries the compound and makes it brittle so it can be scraped or brushed away. Finish with a scouring using household detergent or a TSP solution.

Chewing Gum. Begin by using the technique just described for caulking compounds. Solvents such as chloroform and carbon disulfide may also dissolve gum.

Beverage Stains. Soft drink, alcohol, coffee, and tea stains can be removed by applying a bandage saturated with one part glycerol (glycerin) diluted with four parts water. Two parts of isopropyl alcohol added to the mix will speed up the cleaning process. Methods described for smoke stain removal may also be effective on dark coffee or tea stains.

CONCRETE REPAIR PRODUCTS

A variety of concrete repair products are available for patching and resurfacing work. These products have been specially formulated for specific types of repairs, so read the manufacturer's labeling to match the right product with the job at hand. The following are general descriptions of the most used concrete repair products.

- **Vinyl Resin Patchers.** These patching products blend vinyl resins, very fine sand, and portland cement to form a completely self-bonding material suitable for indoors and outdoors. They can be used to patch cracks and holes, and repair chipped edges on steps, ledges, and cast concrete. Vinyl resin cements can be used to resurface small areas of damaged concrete, but most manufacturers do not suggest using this type of patcher as a resurfacing product for large areas. Their

main advantage is that they can be feathered to a thickness of 1/16".
- **Quick-Setting Cements.** These fast-setting (five to ten minutes) cements are the ideal repair products for chipped concrete edges. As the cements set up, they can be shaved and sculpted to match the original shape of the object. They can also be used to patch holes and cracks when high early strength and setup are needed.
- **Hydraulic Water-Stopping Cements.** These specialty cements are designed to plug cracks and stop the flow of water by extremely rapid (three to five minutes) setup and expansion. See chapter 9 for more details.
- **Caulking.** Although not a cement-based product, acrylic latex caulking is often used to seal small cracks in concrete surfaces that are not exposed to traffic.
- **Resurfacing Products.** Portland cement-based sand mixes and specialty top-bonding mixes are available for larger resurfacing projects for walkways, patios, slab decks, and so on.
- **Surface-Bonding Cements.** These are fiberglass-reinforced cement mixes used to repair and restore wall surfaces only. Application thickness is usually much less than that used with horizontal resurfacing products.

REPAIRING MINOR CRACKS

Minor flaws, such as small cracks, are unsightly and can lead to further deterioration of the concrete if left unrepaired. Vinyl resin patchers are the ideal repair product. After cleaning the crack of all dirt, dust, and loose concrete, the patcher should be applied in layers of 1/4" or less. If more than one layer is needed for the repair, several days of curing time should be allowed between layers. Because vinyl-based patchers may produce a slightly stiff mix, it's important to keep the trowel wet and clean during application. This keeps the patcher from curl-

Repairing a small, deep crack using a vinyl resin patching material

ing up on the trowel. Follow the manufacturer's mixing directions, and resist the temptation to add more water to the mix.

Cracks that are too small to accept the cement-based patcher can be filled with concrete repair caulks. These acrylic latex caulks dry quickly to a tough, flexible finish that blends

Caulking smaller cracks to prevent water penetration

with concrete surfaces and stands up to traffic on walkways and driveways. You should also caulk any areas where concrete and masonry meet the wooden or steel framing members of your house. Just be sure that the gaps are filled so that the material overlaps the edges of the opening.

REPAIRING LARGER CRACKS AND HOLES

For cracks shallower than 1-1/2" to 2", a vinyl resin patcher or quick-setting cement can be used. For cracks deeper than 2", the repair can be made using a prepackaged sand mix or your own mix of one part portland cement to three parts fine sand. Enough water should be added to make the mixture soft and spreadable, like soft ice cream.

Break out all old and crumbling edges of the crack or hole. Undercut the edges of the crack at least 1" below the surface using a small sledgehammer and a cold or brick chisel. The idea is to form a deep, keyed area that will help hold the patch in place.

Brush out all loose concrete and dirt. Use a shop vacuum or hand vacuum to suck out any remaining dust. Manufacturer's directions may or may not suggest dampening the area before applying the patching material. When patching vertical surfaces, apply the patch in thin layers of

Undercutting the edges of the hole to accept the patching material

Clean the area completely before applying the patch.

Pry loose all flaking concrete with the tip of a mason's trowel.

1/4″ or so, waiting for the mix to stiffen slightly between layers. If the material is applied too heavily, its weight may cause it to sag or pull loose.

For larger holes, an application of concrete bonding adhesive to the repair area is recommended.

REPAIRING FLAKING AND CRUMBLING SURFACES

Flaking or chipped concrete is usually the result of poor finishing techniques or frequent exposure to deicers or other chemicals. Small patches of chipped or flaked concrete can be repaired using vinyl patching material as follows:

1. Vigorously rub a wire brush against the crack or flaking area and remove any

Clean away all dirt and loose concrete.

Applying the patch in layers. If the area is subject to water penetration, use a hydraulic water-stopping cement.

cracked or crumbling concrete. You can also use the tip of the trowel to pry up loose sections of the flaked concrete.
2. Use a regular brush to remove loose particles and dust. Clean the area with an all-purpose cleaner for removing oil, grease, or soil buildup from the concrete. Rinse with clean water. Do not leave any surface water.
3. Mix the patching compound according to manufacturer's instructions. Some compounds must stand for several minutes after mixing before they can be

157

Mix the patching compound according to manufacturer's directions.

Apply and featheredge the patching material with a trowel.

Spot resurfacing deeper recesses

applied. Never mix more compound than you can comfortably apply in a half hour.

4. Apply the mix with a trowel and taper the edges smooth with the surrounding concrete. This is called *featheredging*.
5. Deeper spot repairs should be handled like the large crack and hole repair explained earlier. Break away the old concrete to a depth of 1″ or so. Clean the area completely, including any grease or oil spots that may prevent good bonding.

6. Use a sand mix or vinyl patcher. Coat the surface with bonding adhesive to increase the patch's holding power. Sand mix can be placed in a single layer, but vinyl resin patchers should be applied in thin layers. Trowel the top surface to match the surrounding concrete and cure properly.

REPAIRING BROKEN EDGES AND CORNERS

Many homeowners neglect repairs to precast pieces, chipped concrete statuary, and even angled concrete (such as steps and corners) out of a lack of confidence in their abilities to work with concrete, or a feeling that the repairs are not worth the effort of making suitable forms.

Quick-setting cements make these and similar jobs simple enough for even the inexpe-

Preparing broken edges and corners

Shape and shave the quick-setting cement once it begins to set up.

rienced do-it-yourselfer to achieve more than satisfactory results. As mentioned earlier, these cements are specially formulated for use where rapid setting and high strengths are needed. They set in five to ten minutes and can be sculpted with a trowel to match the surrounding surface without any lengthy delays. No forms are needed for small repairs. Proceed as follows:

1. Undercut and/or square the edges of the damaged area with a hammer and chisel. Be sure to wear safety glasses to prevent injury from flying chips of concrete. If, when cleaning out the hole, the hole becomes larger and larger (as it sometimes tends to do), you can punch a hole in the good, firm concrete and drive in a metal pin, and then make the patch. The pin helps hold the concrete in

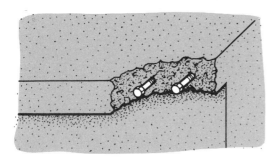

Metal pins inserted in drilled holes can reinforce corner repairs.

place. The pin must be 1/2" lower than the finished surface.
2. Brush all loose material and dirt from the area to be repaired. Dampen the damaged surface or apply a coating of concrete bonding adhesive.
3. Apply the cement to the damaged surface, rough-shaping it to the proper contours as it is applied. Prepare only as much cement as can be applied in about five minutes.
4. After the concrete receives its initial set (several minutes), sculpt it to its finished form by shaving it with the edges of the trowel.

Repairs with Forms

When the entire leading edge of a step tread or a large portion of a corner is damaged or missing, it is best to use some type of simple formwork. The forms will hold the repair material in place as it sets up and help guarantee smooth, clean lines, and sharp edges.

Undercut and widen the broken area as described above. Use a concrete bonding adhesive and sand mix, quick-setting cement, or vinyl patcher. Do not exceed the application thickness stated on the repair product package.

Do not attempt to finish or shape the edges or corners until the patch has been in place long

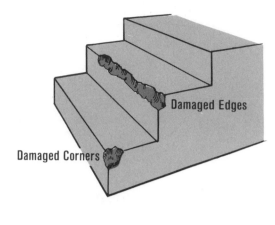

Include Wood Molding in the Forms for More Decorative, Protruding Stair Treads.

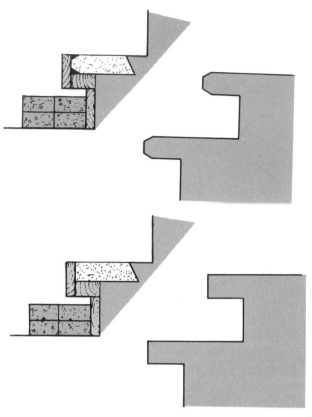

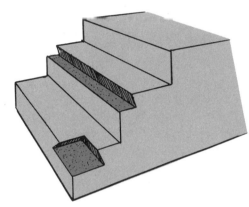

Damaged Edges

Damaged Corners

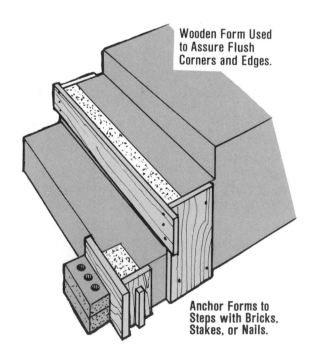

Wooden Form Used to Assure Flush Corners and Edges.

Anchor Forms to Steps with Bricks, Stakes, or Nails.

Repairing damaged edges and corners using formwork

When the entire step edge is restored, moldings and formwork can be used to create decorative treads.

enough to retain a firm thumbprint. You can then finish and smooth the edges and corners flush with the adjoining step surfaces using a trowel. Remember with quick-setting cements this may only take a minute or two, so be prepared.

As shown, including wood moldings in the forms will create more decorative protruding stair treads. Cure the patch properly for best results.

RESURFACING OLD CONCRETE

Badly chipped or pitted sidewalks and slabs are excellent candidates for resurfacing pro-

This badly pitted walk is an ideal candidate for complete resurfacing.

vided there are no major breaks or loose sections. Concrete that has suffered large cracks, fractures, and upheavals should be broken up and replaced. But if the body of the concrete is still sound, consider a facelift with resurfacing materials.

Resurfacing involves troweling a fresh layer of concrete onto the damaged areas. The work involved is considerably less than placing new concrete, and the result can be just as good looking and durable as a new sidewalk or slab.

The resurfacing mix consists of a blend of sand and cement that forms a smooth, easy-to-work top layer. The depth of this top sand mix layer should not exceed 2″. If sections of the work are greater in depth than 2″, fill in these low areas with regular concrete prior to topping with the same mix.

Boards staked along the sides of the existing walk or slab provide the formwork needed for clean edges and a level top surface.

While it's possible to mix the sand, cement, and stone yourself, this project can be easily handled with premixed sand and concrete products. These just-add-water premixes offer sev-

eral convenient advantages. For example, the small stone aggregate used in concrete premixes is just the right size for filling in deep pockets and low areas. With a sand premix you're guaranteed the proper sand-to-cement ratio for the all-important top coat. Premixed materials can be prepared in small batches, and the volume of the mix needed is usually not great enough to be cost-prohibitive. If you prefer to mix the top coat yourself, use a mix ratio of one part portland cement to three parts clean sand.

Success depends on the bond between the new top coat and the old concrete. That's why it's vital to remove all loose concrete, dirt, and plant material from the old surface.

Also take advantage of the specialty products available to increase bonding strength. Brushable concrete bonding adhesive applied to the old surface will penetrate its pores and form a chemical bond with the new topping. Water-resistant acrylic fortifiers can also be added to the sand mix as part of the mix water. These fortifiers help strengthen and seal the new surface. And, of course, proper curing of the resurfaced sidewalk or slab is essential.

Site Preparation

1. Dig a small trench along the edge of the damaged surface so that forms can be set in place level with the old concrete surface.

2. Remove all broken and loose concrete, dirt, dust, and plant remains to form a sound base for the new topping.

3. Set and stake forms against the old sidewalk or slab. Backfill against the forms to ensure sufficient support.

4. Use a level to make sure the forms are set at the correct height and that a slight side slope exists to help with water drainage.

Making the Repair

1. Apply a commercial concrete bonding adhesive to the clean surface as you would a thick coat of paint. Wait until the bonder becomes tacky before placing the concrete or sand mixture.

2. When needed, fill in low spots with regular concrete mix, building up to a point 1″ to 2″ below the top of the forms. Be sure to work in the mix along the sides of the forms.

3. Blend the sand mix with water until a plastic-like consistency is obtained. Any liquid fortifiers added to the mix are part of the mix "water" and should be added first before mixing begins.

4. Apply the sand mix over the top of the damaged concrete and any fresh concrete used to build up low spots. The sand topping should be level with the top of the forms. Taper out the topping onto the adjacent sound concrete to a thickness of about 1/4".

5. Screeding is not usually needed. Simply trowel the surface smooth using a steel finishing trowel. Edge using a concrete edger, if desired.

6. For a nonskid surface, apply a broomed finish. Always pull the broom toward you, use light pressure, and do not overlap the strokes.

7. Cure the new topping as you would a new concrete installation. To prevent discoloration of the slab when using plastic sheeting, the plastic must be kept flat on the concrete surface.

REPLACING DAMAGED CONCRETE SLABS

When cracks and crumbled areas extend all the way through the slab in numerous locations, there is not a strong enough base for resurfacing. The best repair option is to break up the damaged area completely, remove it, and place new concrete.

The repair will look neater and probably hold up longer if you replace all concrete between a given set of control joints in the walk. Use a brick chisel and small sledgehammer at the control joint to break the damaged section away from the sound concrete. Then use a regular sledgehammer to break the concrete into small sections for removal. If you can, try to elevate the damaged slab slightly off the ground so that it is easier to break with the sledgehammer.

Once the old, damaged section is removed, the repair becomes more or less a standard walk building project. As in the resurfacing project outlined above, stake formwork in line with the existing walkway. Be certain that the replacement section will be as thick as the existing walk. Small pieces of the broken-up walk can be used as a gravel base. Install expansion joint fiber strips where the new section meets existing concrete. Rebar inserted into holes drilled into the existing concrete will help tie the sections together and prevent shifting and settling.

Place and finish the concrete as described in chapter 5.

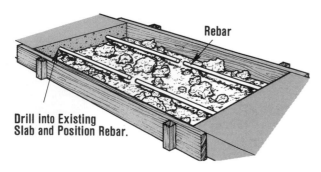

Rebar

Drill into Existing Slab and Position Rebar.

Details for complete replacement of a damaged walk section

REPOSITIONING SETTLED AND SHIFTED SLABS

Walkways may shift, settle, and heave due to changes in their subgrade or pressures from growing tree roots. If the walk is built correctly, this cracking and movement may only occur at control and expansion joints, with complete sections of the walk remaining intact.

If this is the case, the most economical repair method is to reposition the slab sections on a new base. If the walk has not cracked at its control joints, use a chisel and small sledgehammer to break it cleanly at these points. Repositioning slabs of any size will require the aid of a helper. Wear heavy work gloves and be careful not to pinch fingers or toes under the slab edge.

Begin by determining the desired finished height for the repositioned sections. String a mason's line or stake a 2×4 piece of lumber at this height to give yourself a visual reference.

Use a long iron pry bar and small block of wood to create a fulcrum for lifting leverage. Never rest the pry bar directly on the adjacent concrete slab to use as a lift. The edge may chip or crumble under the pressure. Pickaxes can also be used as levers.

Flat stones or pieces of brick should be used to shim the slab to the correct level. Push these in position with a rake or board. Never put your

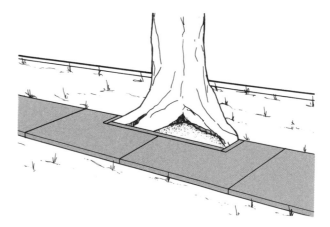

Walks damaged by tree roots should be broken up and replaced with a narrower section. Use pressure-treated lumber as an inside border.

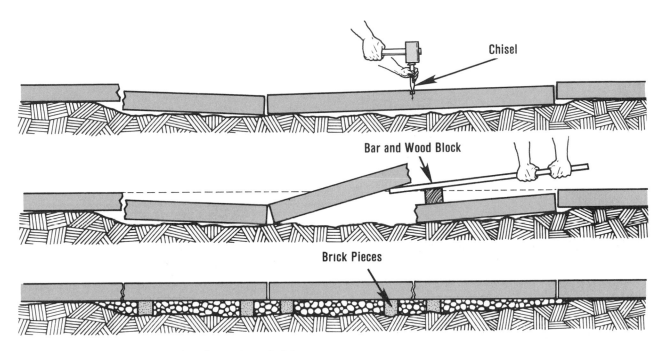

Details for releveling shifted and misaligned slabs

hand or arm beneath the lifted slab. Shim to a height slightly higher than required to allow for anticipated settlement. Once the slab is firmly supported on the bricks or stones, pack the voids beneath the slab with pea gravel.

If intruding tree roots are the cause of the problem, they can be cut away and removed. Cutting away too much of a tree's root system may kill the tree, so discretion is advised. A safer treatment is to narrow the walk at this point as shown. This will solve the problem for several years until the growing root system again infringes on the walk.

RESTORATION WITH SURFACE-BONDING CEMENTS

Mortarless block construction with special premixed surface-bonding cements is discussed in detail in chapter 15. These cement-based blends of reinforcing fiber and fine sand can also be used to renovate existing concrete block, brick, stone, concrete, and other structurally sound walls.

Surface-bonding cements are particularly effective in restoring older buildings whose walls have begun to crack and loosen from the effects of time. A 1/8″ coating of this cement creates an attractive stucco-like finish that fills small cracks and imperfections. The coating also seals the wall against air and water infiltration.

Because surface-bonding cements are normally highly water-resistant, they can be used as a lining on cisterns, holding ponds, and other water-holding vessels. Just make absolutely sure the product you select is approved for use with potable water so it is safe for both human and animal use.

All surfaces should be clean and dry to ensure maximum bonding of the surface-bonding cement. Remove all flaking paint, dirt, and loose concrete. Patch large cracks and holes using the methods outlined earlier in this chapter. Use hydraulic cement to patch cisterns and tanks. Apply the surface-bonding cement directly to the wall using a finishing trowel and an upward sweeping motion. See chapter 15 for more details on working with surface-bonding cements.

Chapter 11

MASONRY MATERIALS

Masonry refers to any type of construction that involves the laying of building units, such as brick, stone, or concrete block, with or without the use of a cementing agent to hold them together. When a cementing agent is used in masonry work, it's known as *mortar*.

The art of masonry is thousands of years old. Ancient peoples bonded together sun-dried or fired brick with whatever "mortar" was handy. Assyrians and Babylonians used a clay-based mortar, while Egyptians favored a mix of limestone and gypsum. Roman masons benefited from the use of their naturally blended hydraulic cements, but throughout most of history bricklayers were forced to rely on mixes primarily made up of sand and lime. Unfortunately, these crude mortars required months and even years to harden. The brick or stone had to be carefully fitted together, and the thin mortar joints acted more as caulking than a true cement bond.

The development of portland cement by Joseph Aspdin revolutionized masonry construction. In fact, Aspdin's experiments were triggered by his desire to find a better bricklaying mortar for lighthouse construction. Adding portland cement to the customary mixture of sand and lime greatly strengthened the mortar and reduced its setup time from days to hours. With this mixture, masons could lay up brick faster and use thicker mortar joints to help correct small imperfections between units.

MODERN MORTARS

Mortar for laying up brick or concrete block consists of three major components: portland

TABLE 11-1: PROPORTIONS FOR MIXING MORTAR

Type	Parts by Volume			Strength
	Portland Cement	Hydrated Lime	Sand (maximum)	
M	1	1/4	3-3/4	High
S	1	1/2	4-1/2	High
N	1	1	6	Medium
O	1	2	9	Low

cement, hydrated lime, and sand. The lime makes the mortar tacky and easier to work, and the sand adds volume. When mixed with the proper amount of water, the three ingredients form a somewhat plastic, sticky mortar.

Contractors often mix their own mortar from scratch using correct proportions of cement, lime, and mortar grade sand. If you decide to save money and go this route, use table 11-1 to determine how much of each ingredient to use. Mortar sand has finer particles than the coarse sands used in concrete. Always mix the ingredients with clean drinking water, using as much as you need to make the mortar workable.

Mortar is generally divided into four types: M, S, N, and O. Type M is for general use where high strength is required; it's also ideal in areas susceptible to severe frost. Type S is best where high lateral strength is needed, such as exterior walls; type N should be used for load-bearing walls; and type O for nonload-bearing walls.

Another option is to purchase mortar cement, which is a mixture of portland cement and lime. All that's needed is to add the sand and water. For really simple mixing, you might want to buy ready-mixed mortar. This premix contains all three ingredients; just add the water and

The mortarboard and wheelbarrow are important tools for mixing mortar.

it's ready to use. Regardless of the type of mortar you use, it's very important that the ingredients are thoroughly mixed.

For very small jobs, the mortar can be mixed on a mortarboard. Otherwise, you should use a wheelbarrow or power mixer. Blend the ingredients while they're dry then add water gradually while continuing to mix. Since the cement, lime, and sand are different colors, a good sign that the mixing has been successful is when the mortar takes on a uniform look. A good mix holds together and spreads easily.

Never mix more mortar than you can use in about two hours. Don't let mortar sit for too long after it has been mixed, because it has a tendency to dry out and stiffen; use it as soon as possible. It's also not a good idea to add lots of water to a batch of mortar that has been sitting around for more than two hours in order to make it workable again. If you must retemper mortar, use the minimum amount of water needed to do the job and be sure to mix thoroughly.

GROUT

Grout is an essential element of reinforced concrete block masonry. In reinforced load-bearing wall construction, grout is usually placed only in those wall spaces containing steel reinforcement such as bond beams, pilasters, or piers. Grout is also used to secure various types of wall ties and anchor bolts to block units.

The grout bonds the block and steel so that they act together to resist loads. In some reinforced load-bearing walls, all cores—with and without reinforcement—are grouted to further increase the wall resistance to loads. Grout is sometimes used in nonreinforced load-bearing

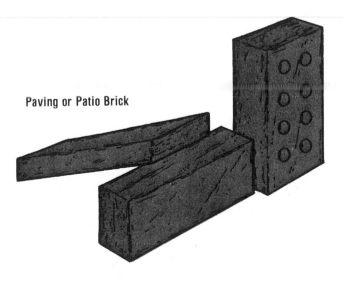

Paving or Patio Brick

Cored Brick

Common or Building Brick

Brick types

TABLE 11–2: GROUT PROPORTIONS*

Type	Portland Cement	Hydrated Lime or Lime Putty	Aggregate Measured in a Damp, Loose Condition	
			Fine	Coarse
Fine grout	1	0 to 1/10	2-1/4 to 3 times the sum of the volumes of the cementious materials	—
Coarse grout	1	0 to 1/10	2-1/4 to 3 times the sum of the volumes of the cementious materials	1 to 2 times the sum of the volumes of the cementious materials

*The sum of the volumes of the fine and coarse aggregate should not exceed four times the sum of the separate volumes of the cement and lime used.

block wall construction to give added strength. This is accomplished by filling a portion or all of the cores.

Mortar is not grout. Grout can be produced by adding more water to mortar that has a limited amount of lime (table 11–2).

TYPES OF BRICK

For a combination of durability and attractiveness, brick can't be beat. Most bricks are manufactured by firing molded clay or shale, and vary greatly in color, size, and texture. All brick sold by reputable building supply dealers must meet the standards established for the trade. It's convenient to place all brick types into one of four general categories, as follows:

Face brick is probably the highest quality available. The texture and color are uniform, and it's free from flaking, chipping, cracking, warpage, and other such defects. Since face brick is so nearly perfect, it's commonly used on the exposed face of walls. Common face brick shades include brown, red, gray, yellow, and white.

Common brick, also known as building brick, is as strong as the face type. However, it contains more imperfections; for example, the color and texture aren't as carefully controlled. Common brick has no special scorings or mark-

ings, and is normally used for general building purposes and for backing courses in walls.

Fire brick is made from a special clay that enables it to withstand extremely high temperatures. This makes it ideal for use in fireplaces, barbecues, and stove liners. Fire brick is yellow in color, often hand molded, and larger than other types of brick.

Paving brick, also known as patio brick, is sized for use with or without mortar joints. It's baked for a long period and, like fire brick, is made from special clay. Because paving brick is very hard and resists cracking even under heavy loads, it's ideal for building patios and walks.

Any of the above types of brick can be cored and/or ceramic faced. Coring helps reduce the overall weight of a brick without affecting its strength. Cored bricks are most often used where the holes won't be visible, such as walls and planter boxes. A ceramic coating on the face of a brick allows for easy cleaning in interior applications.

Brick Grades

Its durability notwithstanding, brick can still be damaged by freezing weather. The ability of a brick to resist such conditions is directly related to the amount of moisture inside it. Longer baking times reduce the water absorption capacity of brick, thus making it less susceptible to cold weather. The three grades of brick are:

- SW, or severe weathering, offers the highest resistance to freeze–thaw and rain–freeze conditions. It's recommended in areas with subzero winters and for any project in which the brick contacts the ground, such as retaining walls.
- MW, or moderate weathering, can stand rain–freeze conditions, but not severe ones. Use it in areas with subfreezing weather, but not the bitter and extensive cold found in the northern United States and Canada.
- NW, or no weathering, is intended mainly for interior projects. Use it outdoors only in very mild climates where even hard frost isn't a factor.

Brick Sizes

Since most bricks aren't made to exacting specifications, they're usually referred to by nominal dimensions rather than their actual size, the same as lumber. Just as the standard 2×4 is really 1-1/2″×3-1/2″, bricks are also a bit smaller than their listed dimensions. The nominal size of a brick is simply its actual size plus the width of the mortar joint used. Brick is normally laid up using 3/8″ or 1/2″ mortar joints, with the latter being most popular.

When buying brick, specify the nominal size plus the mortar joint thickness you will be using. The nominal size of common brick is stated as 2-2/3″×4″×8″. For lay up with 3/8″ joints, the actual size of common brick manufactured is 2-1/4″×3-5/8″×7-5/8″. The actual size of common brick manufactured for use with 1/2″ joints is 2-1/4″×3-1/2″×7-1/2″. Notice that the 2-2/3″ nominal height measurement is about halfway between the 2-5/8″ and 2-3/4″ actual heights per course created with 3/8″ and 1/2″ mortar joints, respectively.

Table 11–3 lists the nominal dimensions of popular brick types plus their actual dimensions for use with 3/8″ and 1/2″ mortar joints. Tables 11–4 and 11–5 list the nominal wall heights created by courses using common brick and 1/2″ and 3/8″ mortar joints.

TABLE 11–3:   DIMENSIONS OF COMMONLY USED BRICK (IN INCHES)

Brick Type	Nominal Dimensions H×W×L	Mortar Joint Thickness	Actual Dimensions H×W×L
Common	2-2/3×4×8	3/8	2-1/4×3-5/8×7-5/8
		1/2	2-1/4×3-1/2×7-1/2
Norman	2-2/3×4×12	3/8	2-1/4×3-5/8×11-5/8
		1/2	2-1/4×3-1/2×11-1/2
Roman	2×4×12	3/8	1-5/8×3-5/8×11-5/8
		1/2	1-1/2×3-1/2×11-1/2
Baby Roman	2×4×8	3/8	1-5/8×3-5/8×7-5/8
		1/2	1-1/2×3-1/2×7-1/2
SCR	2-2/3×6×12	3/8	2-1/4×5-5/8×11-5/8
		1/2	2-1/4×5-1/2×11-1/2
Economy	4×4×12	3/8	3-5/8×3-5/8×11-5/8
		1/2	3-1/2×3-1/2×11-1/2
Jumbo Six	4×6×12	3/8	3-5/8×5-5/8×11-5/8
		1/2	3-1/2×5-1/2×11-1/2
Jumbo Eight	4×8×12	3/8	3-5/8×7-5/8×11-5/8
		1/2	3-1/2×7-1/2×11-1/2

TABLE 11-4:	NOMINAL HEIGHTS OF COMMON BRICK WALLS USING 1/2" MORTAR JOINTS		
Courses	Height	Courses	Height
1	0'–2-3/4"	31	7'–1-1/4"
2	0'–5-1/2"	32	7'–4"
3	0'–8-1/4"	33	7'–6-3/4"
4	0'–11"	34	7'–9-1/2"
5	1'–1-3/4"	35	8'–0-1/4"
6	1'–4-1/2"	36	8'–3"
7	1'–7-1/4"	37	8'–5-3/4"
8	1'–10"	38	8'–8-1/2"
9	2'–0-3/4"	39	8'–11-1/4"
10	2'–3-1/2"	40	9'–2"
11	2'–6-1/4"	41	9'–4-3/4"
12	2'–9"	42	9'–7-1/2"
13	2'–11-3/4"	43	9'–10-1/4"
14	3'–2-1/2"	44	10'–1"
15	3'–5-1/4"	45	10'–3-3/4"
16	3'–8"	46	10'–6-1/2"
17	3'–10-3/4"	47	10'–9-1/4"
18	4'–1-1/2"	48	11'–0"
19	4'–4-1/4"	49	11'–2-3/4"
20	4'–7"	50	11'–5-1/2"
21	4'–9-3/4"	51	11'–8-1/4"
22	5'–0-1/2"	52	11'–11"
23	5'–3-1/4"	53	12'–1-3/4"
24	5'–6"	54	12'–4-1/2"
25	5'–8-3/4"	55	12'–7-1/4"
26	5'–11-1/2"	56	12'–10"
27	6'–2-1/4"	57	13'–0-3/4"
28	6'–5"	58	13'–3-1/2"
29	6'–7-3/4"	59	13'–6-1/4"
30	6'–10-1/2"	60	13'–9"

TABLE 11-5:	NOMINAL HEIGHTS OF COMMON BRICK WALLS USING 3/8" MORTAR JOINTS		
Number of Courses	Vertical Height	Number of Courses	Vertical Height
1	2-5/8"	31	6'–9-3/8"
2	5-1/4"	32	7'–0"
3	7-7/8"	33	7'–2-5/8"
4	10-1/2"	34	7'–5-1/4"
5	1'–1-1/8"	35	7'–7-7/8"
6	1'–3-3/4"	36	7'–10-1/2"
7	1'–6-3/8"	37	8'–1-1/8"
8	1'–9"	38	8'–3-3/4"
9	1'–11-5/8"	39	8'–6-3/8"
10	2'–2-1/4"	40	8'–9"
11	2'–4-7/8"	41	8'–11-5/8"
12	2'–7-1/2"	42	9'–2-1/4"
13	2'–10-1/8"	43	9'–4-7/8"
14	3'–0-3/4"	44	9'–7-1/2"
15	3'–3-3/8"	45	9'–10-1/8"
16	3'–6"	46	10'–0-3/4"
17	3'–8-5/8"	47	10'–3-3/8"
18	3'–11-1/4"	48	10'–6"
19	4'–1-7/8"	49	10'–8-5/8"
20	4'–4-1/2"	50	10'–11-1/4"
21	4'–7-1/8"	51	11'–1-7/8"
22	4'–9-3/4"	52	11'–4-1/2"
23	5'–0-3/8"	53	11'–7-1/8"
24	5'–3"	54	11'–9-3/4"
25	5'–5-5/8"	55	12'–0-3/8"
26	5'–8-1/4"	56	12'–3"
27	5'–10-7/8"	57	12'–5-5/8"
28	6'–1-1/2"	58	12'–8-1/4"
29	6'–4-1/8"	59	12'–10-7/8"
30	6'–6-3/4"	60	13'–1-1/2"

Buying Brick

To select the best brick for a particular project, visit a supplier's showroom and inspect the samples on hand. You'll be able to compare a variety of colors and textures; the display will also have simulated mortar joints on a panel or wall section, with the joints struck in popular finishes. When you've made your choice, use the manufacturer's identification number (this might also be called the range or blend number) to place your order. Keep in mind that your choice might not be immediately available, since some brick kiln runs are made only at certain times of the year.

Bricks can be bought individually, in cubes of five hundred, or by the thousand. When you order, make sure you and the dealer are speaking the same language. As pointed out earlier, nominal sizes make allowances for mortar joints, but actual sizes don't. If possible, you might want to consider hauling your own bricks. Dealer delivery charges can be high; in fact, for a small job the delivery can cost as much as the bricks themselves. Be careful not to overload your truck when hauling brick, since a cube of five hundred weighs about a ton.

Make sure you find out from the supplier if additional bricks of the same type will be available in the future, in case you run out or decide to expand the project. Some bricks are made on a limited basis and then discontinued. Also, if you're using cored brick, make sure that solid-end brick is available for the ends of walls, windowsills, and so forth.

If you're considering buying used brick, keep in mind that it is not always structurally sound. It can be so filled with impossible-to-remove mortar that a fresh mortar joint will be only half as strong as it should be. And bricks manufactured a generation ago simply are not as reliable as those made today. For these reasons, most professionals limit their use of used brick to decorative purposes and veneering. The risk is just too great to use it for projects that must sustain loads.

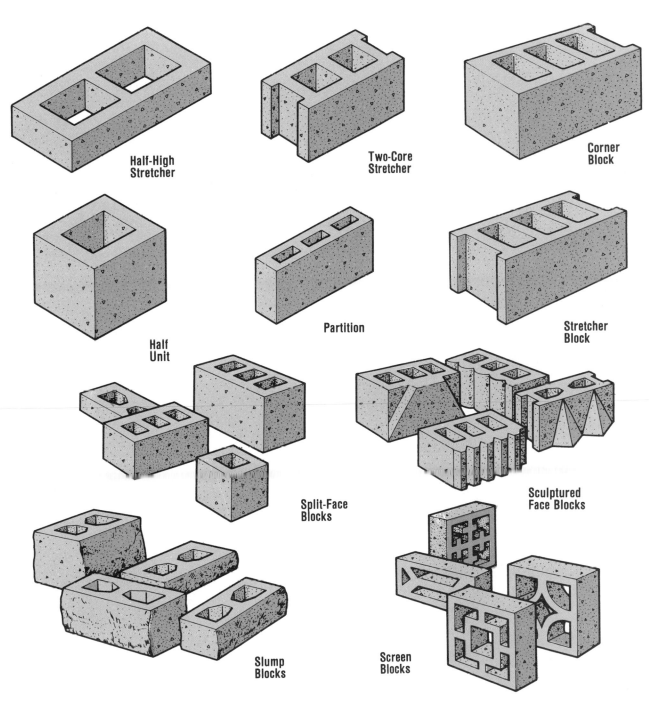

Half-High Stretcher

Two-Core Stretcher

Corner Block

Half Unit

Partition

Stretcher Block

Split-Face Blocks

Sculptured Face Blocks

Slump Blocks

Screen Blocks

Variations of concrete block

CONCRETE BLOCK

Concrete block is ideally suited for foundation walls, retaining walls, and other jobs where strength is important. It costs less and takes less time to lay than bricks, so it's a good choice for the beginner. Also, the appearance of concrete block can be enhanced in a variety of ways. It can be faced with more appealing brick or stone units, or given a stucco or surface-bonding cement finish. Concrete block can also be coated with compatible paints, masonry coatings, or pigmented water sealers.

Concrete block units are manufactured in full and half sizes. As with brick, nominal dimensions are used; the standard 8×8×16 block is actually 7-5/8"×7-5/8"×11-5/8". This allows for a 3/8" mortar joint.

Concrete blocks weigh anywhere from 25 to 50 pounds, depending on the aggregate they contain. Common aggregates include gravel, crushed stone, sand, and blast furnace slag. Lightweight blocks, or cinder blocks, are also available. These contain clay, slate, expanded shale, or cinder as their aggregate, and thus are a lot lighter and easier to work with than standard block. Keep in mind that most building codes call for load-bearing block for foundation walls.

Many different modular sizes of concrete block are produced in addition to the standard type. Some resemble oversized brick and can be either solid or cored, while others are partially

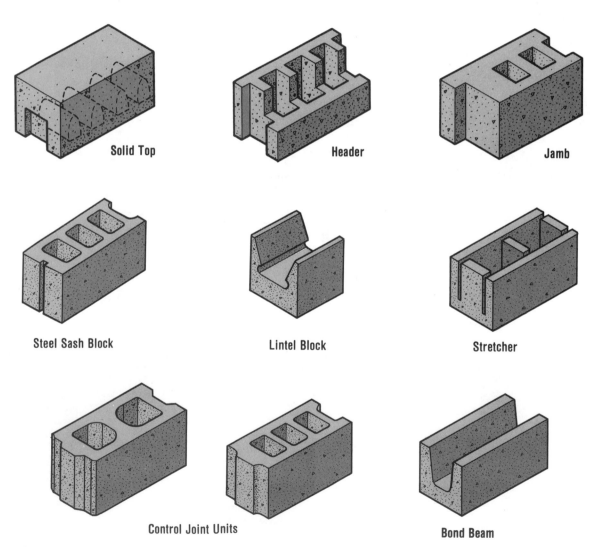

Solid Top   Header   Jamb

Steel Sash Block   Lintel Block   Stretcher

Control Joint Units   Bond Beam

Specialty concrete block

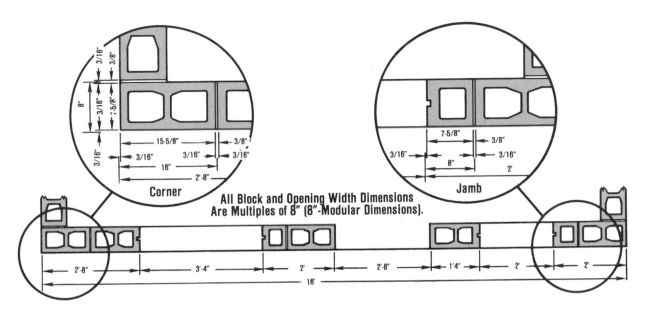

All Block and Opening Width Dimensions
Are Multiples of 8″ (8″-Modular Dimensions).

Modular planning of a wall

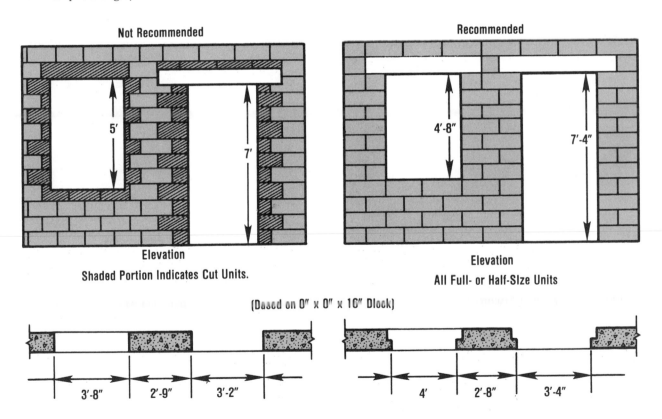

Modular planning cuts down on work time and material waste.

faceted to form a pattern when several are used together. Stretcher blocks, commonly used for house foundations, have mortar joint projections at both ends for easy assembly. Split-face blocks feature a rustic facing; slump blocks have a rugged, hand-made look. There are also specialty blocks for working corners, jambs, wall caps, and so on.

Screen blocks are popular indoors and out. They're good for building dividers, and as a patio

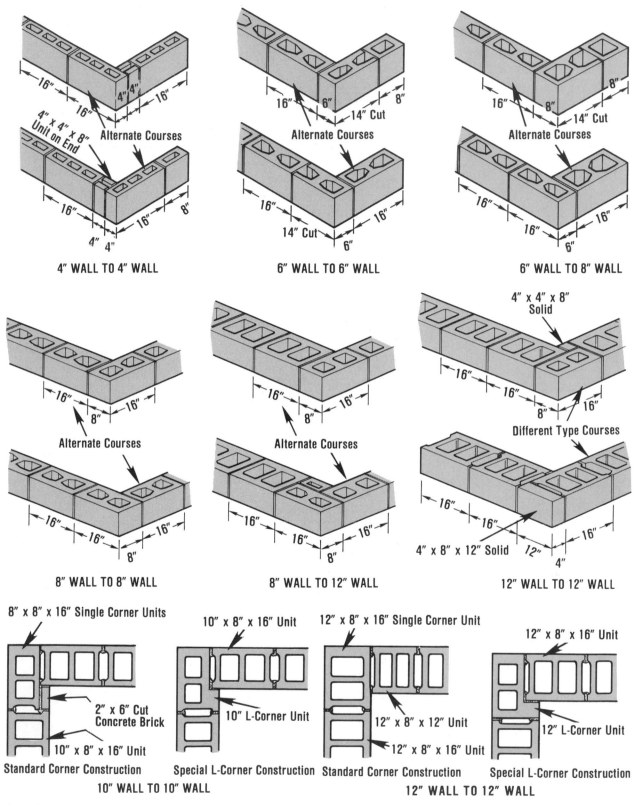

4" WALL TO 4" WALL

6" WALL TO 6" WALL

6" WALL TO 8" WALL

8" WALL TO 8" WALL

8" WALL TO 12" WALL

12" WALL TO 12" WALL

Standard Corner Construction
Special L-Corner Construction
10" WALL TO 10" WALL

Standard Corner Construction
Special L-Corner Construction
12" WALL TO 12" WALL

Standard corner layouts for walls of varying thickness

screen they provide privacy with minimum air blockage. Screen blocks lend a light, airy touch to even large projects. But never use this type of block as the main load-bearing part of a structure.

MODULAR PLANNING

When planning the length or height of brick or block walls, assign multiples of nominal dimensions to avoid having to cut or use half-height units. The width and height of wall openings and the areas between the doors, windows, and corners should also be planned carefully to use standard full-size and half-size units. With modular planning, the work will not only be easier, but the end result will look more uniform and pleasing to the eye.

An example of modular planning with concrete block is to set up the length and width of all walls, doors, and window openings in 8″ increments. This will greatly reduce the amount of block you'll have to cut and fit on the job. Though you might think there's very little difference between a 7′-high door opening and a 7′-4″-high door opening, there really is; the latter makes it possible to use standard full- and half-size blocks, while the 7′ opening means lots of time-consuming cutting. Nonmodular lengths and widths take simply too much extra work to accomplish.

An important consideration in modular planning is the method used for constructing corners. While 8″ thick walls pose no problem, thicker or thinner walls need special attention if you're trying to preserve the 8″ module. On the opposite page are several examples of corner layouts for walls of varying thicknesses.

To find the number of courses and the number of brick or block in each course, simply divide the proposed wall length and height by the nominal dimension of the brick or block, and then vary the figures until you can divide the number by whole units vertically and by whole or half units horizontally. Naturally, if you are dry-stacking the block for mortarless wall con-

	TABLE 11-6: NOMINAL HEIGHT OF CONCRETE MASONRY WALLS BY COURSES	
	Nominal Height of Concrete Masonry Walls	
No. of Courses	**Units 7-5/8″ High and 3/8″ Thick Bed Joint**	**Units 3-5/8″ High and 3/8″ Thick Bed Joint**
1	8″	4″
2	1′4″	8″
3	2′0″	1′0″
4	2′8″	1′4″
5	3′4″	1′8″
6	4′0″	2′0″
7	4′8″	2′4″
8	5′4″	2′8″
9	6′0″	3′0″
10	6′8″	3′4″
15	10′0″	5′0″
20	13′4″	6′8″
25	16′8″	8′4″
30	20′0″	10′0″
35	23′4″	11′8″
40	26′8″	13′4″
45	30′0″	15′0″
50	33′4″	16′8″

(For concrete masonry units 7-5/8″ and 3-5/8″ in height laid with 3/8″ mortar joints. Height is measured from center to center of mortar joints.)

struction with surface-bonding cement, you must use the actual block dimensions when planning the layout. Tables 11–3 through 11–8 will help you estimate various wall heights and lengths when using standard-sized brick and block with typical joint sizes and dry-stack block arrangements.

BRICKWORK AND BLOCKWORK TOOLS

One of the most useful and versatile tools for brickwork and blockwork is the bricklayer's trowel. It's used to mix mortar, to pick it up from the mortarboard, to place and spread it, and to

TABLE 11-7: NOMINAL LENGTH OF CONCRETE MASONRY WALLS BY STRETCHERS

Number of Stretchers	Units 15-5/8" Long and Half Units 7-5/8" Long with 3/8" Thick Head Joints	Units 11-5/8" Long and Half Units 5-5/8" Long with 3/8" Thick Head Joints
	Nominal Length of Concrete Masonry Walls	
1	1'-4"	1'-0"
1-1/2	2'-0"	1'-6"
2	2'-8"	2'-0"
2-1/2	3'-4"	2'-6"
3	4'-0"	3'-0"
3-1/2	4'-8"	3'-6"
4	5'-4"	4'-0"
4-1/2	6'-0"	4'-6"
5	6'-8"	5'-0"
5-1/2	7'-4"	5'-6"
6	8'-0"	6'-0"
6-1/2	8'-8"	6'-6"
7	9'-4"	7'-0"
7-1/2	10'-0"	7'-6"
8	10'-8"	8'-0"
8-1/2	11'-4"	8'-6"
9	12'-0"	9'-0"
9-1/2	12'-8"	9'-6"
10	13'-4"	10'-0"
10-1/2	14'-0"	10'-6"
11	14'-8"	11'-0"
11-1/2	15'-4"	11'-6"
12	16'-0"	12'-0"
12-1/2	16'-8"	12'-6"
13	17'-4"	13'-0"
13-1/2	18'-0"	13'-6"
14	18'-8"	14'-0"
14-1/2	19'-4"	14'-6"
15	20'-0"	15'-0"
20	26'-8"	20'-0"

(Actual length of wall is measured from outside edge to outside edge of units and is equal to the nominal length minus 3/8" [one mortar joint].)

TABLE 11-8: DIMENSIONS OF DRY-STACKED CONCRETE BLOCKS (STANDARD BLOCK UNITS 7-5/8"×7-5/8"×15-5/8")

Number of Blocks	Length (Laid End to End)	Height* (Stacked)
1	1'-3-5/8"	0'-8"
2	2'-7-1/4"	1'-3-5/8"
3	3'-10-7/8"	1'-11-1/4"
4	5'-2-1/2"	2'-6-7/8"
5	6'-6-1/8"	3'-2-1/2"
6	7'-9-3/4"	3'-10-1/8"
7	9'-1-3/8"	4'-5-3/4"
8	10'-5"	5'-1-3/8"
9	11'-8-5/8"	5'-9"
10	13'-1/4"	6'-4-5/8"
12	15'-7-1/2"	7'-7-7/8"
15	19'-6-3/8"	9'-6-3/4"

*Includes 3/8" mortar bed, first course only.

good deal of mortar farther from your wrist; this can be very tiring and make the job difficult.

The largest bricklayer's trowel is 9" to 11" long and 4" to 8" wide. A good rule of thumb is to choose the largest trowel that you can use comfortably. Smaller pointing trowels are used for masonry repairs and for redoing damaged mortar joints, a process known as *tuckpointing*. The joint filler has a narrow blade for pushing mortar in cracks.

You'll need a wooden mortarboard or hawk for holding the mortar while you work. The board is often grooved to keep the mortar from sliding off. To prevent the wood from absorbing the moisture from the mortar and drying it out, always wet down the board thoroughly before beginning.

The mortar should be kept rounded up in the center of the mortarboard or hawk, with the outer edges remaining clean. Proper consistency must be maintained at all times. If the layer of mortar spread over the board is too thin, the mortar will dry out quickly; there will also be a greater chance of lumps forming.

A brick hammer is a very versatile tool. The chisel end is used to smooth and shape cut

tap the brick or block down into the mortar bed. It's important to use a trowel that has the right "feel" to it. Those with short, wide blades are better for the beginner. Longer trowels hold a

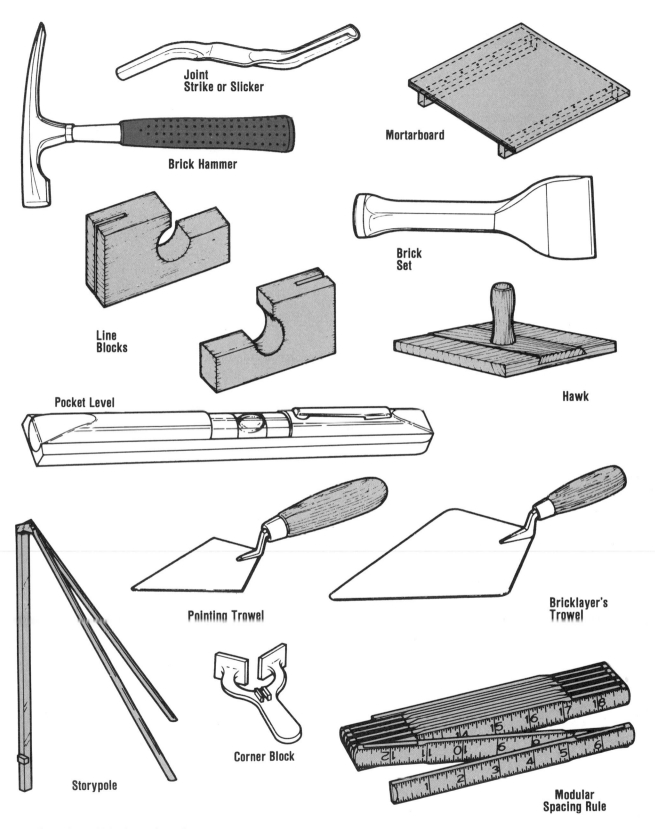

Joint
Strike or Slicker

Mortarboard

Brick Hammer

Brick
Set

Line
Blocks

Hawk

Pocket Level

Pointing Trowel

Bricklayer's
Trowel

Storypole

Corner Block

Modular
Spacing Rule

Brickwork and blockwork tools

bricks, and the square end can break bricks, strike brick sets, and drive nails. The 24-ounce brick hammer is a good choice. Make sure the head is fitted tightly to the handle, and always wear safety goggles.

Brick sets, also known as blocking or mason's chisels, are made of a single piece of steel. They are usually 7″ to 8″ long, with a 3″- to 4″-wide blade that is beveled to the required cutting edge. Use the brick set to make sharp cuts in bricks, or to score bricks for breaking with a hammer. It's a good idea to periodically dress the blade on a grinder to maintain the cutting edge.

Line blocks are used to run a line between brick or block leads so that each course is placed accurately. Modular spacing rules extend to 6′ and are marked on one side to call out course heights. They can be used on their own or to mark off story poles. Both the modular spacing rule and the story pole are helpful in checking the progress of brickwork and blockwork.

A corner block is used to hold a line taut for measuring. Pocket levels come in handy for making checks in confined areas or even on individual bricks. In addition, the layout and leveling tools mentioned earlier are needed for brickwork and blockwork.

ESTIMATING BRICK AND BLOCK NEEDS

Finding the number of bricks or concrete blocks needed to build a wall is easier than you might think. Just multiply the number of units in each course, or row, by the total number of courses. Be sure to subtract the appropriate amounts for windows, doors, and other openings. If you're building a double wythe (two brick wide) wall, just multiply this number by 2. Of course, final count adjustments will be necessary if special bond patterns are used, such as the bull headers (cross tie bricks) used to hold double wythe walls together. See chapter 12 for a full explanation of bricklaying terms and popular bond patterns.

When working with concrete block, you must determine the exact amount of corner, capping, and other specialty blocks you'll need. It's a good idea to roughly sketch out your plans to scale on graph paper. You'll be surprised how helpful this can be for visualizing the project, estimating materials and supplies, and heading off potential trouble areas before the actual construction begins.

Chapter 12

WORKING WITH BRICK

One of the advantages of brickwork is that you don't have to be an expert to achieve beautiful results. Proceeding slowly and carefully, even a beginner can do attractive and lasting work. From curved walks to ornate entryways, the possibilities are endless.

BRICKLAYING TERMINOLOGY

Before beginning any bricklaying project, study the following terms and their definitions. This will help you understand the various brick positions and patterns, as well as the typical mortar joints used.

Bull Header. A rowlock brick laid with its longest dimension perpendicular to the face of the wall.

Bull Stretcher. A rowlock brick laid with its longest dimension parallel to the face of the wall.

Course. A continuous horizontal row of masonry that, bonded together with other courses, forms a masonry structure.

Header. A masonry unit laid flat with its longest dimension perpendicular to the face of the wall. Headers are generally used to tie two wythes of masonry together.

Rowlock. A brick laid on its face, or edge.

Soldier. A brick laid on its end so that its longest dimension is parallel to the vertical axis of the face of the wall.

Stretcher. A masonry unit laid flat with its longest dimension parallel to the face of the wall.

Wythe. A continuous vertical section or thickness of masonry 4" or greater.

BOND PATTERNS

There are a number of bond patterns from which to choose when bricklaying. By using the standard directions for laying common bond, you can use any of these patterns to give variety to your work.

Running Bond. The simplest pattern, this consists of only stretchers. Reinforcing ties are usually used because of the absence of headers. Running bond is common in brick veneer walls and wall cavity construction.

Common or American Bond. As detailed in the step-by-step instructions given later in this chapter, this is a variation of the running bond. It includes a course of full-length headers placed at regular intervals for structural bonding.

Flemish Bond. This pattern uses alternate stretchers and headers, with the headers in alternate courses centered over the stretchers in the intervening courses.

English Bond. This pattern also uses alternate stretchers and headers, but the headers are centered on the stretchers and the joints between the stretchers. The head joints be-

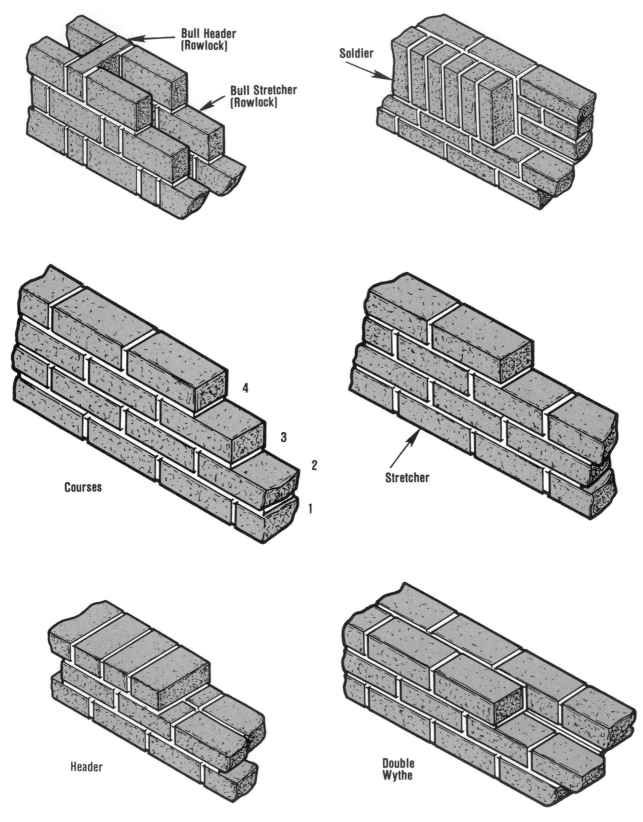

Bull Header (Rowlock)

Bull Stretcher (Rowlock)

Soldier

Courses

4

3

2

1

Stretcher

Header

Double Wythe

Bricklaying terms

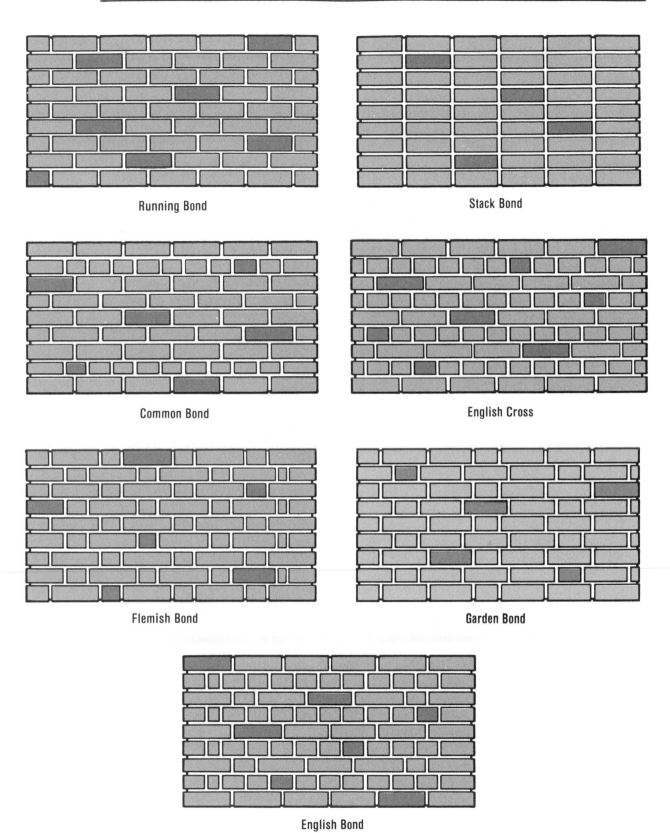

Running Bond

Stack Bond

Common Bond

English Cross

Flemish Bond

Garden Bond

English Bond

Various bond patterns

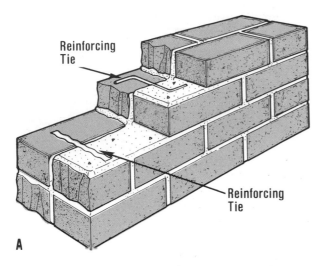

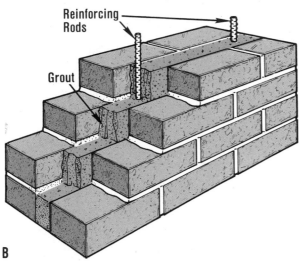

Reinforcement for running bond brickwork: (A) metal ties set in mortar and (B) vertical rebar rods set in grouted cavity between wythes.

tween the stretchers in all the courses line up vertically.

English Cross or Dutch Bond. This is a variation on the English, the only difference being that the vertical joints between the stretchers in alternate courses don't line up vertically. Instead, these joints center on the stretchers themselves.

Stack or Block Bond. This is a weak bond, used normally for decorative effect on veneers. All the vertical joints are aligned, and steel reinforcing ties must be installed if the pattern is being used structurally.

Garden Bond. Each course in this pattern has three consecutive stretchers followed by a header.

BUILDING A WALL USING THE COMMON BOND PATTERN

Whatever brick projects you're considering, they're all basically walls of various configurations. If you can master the common bond bricklaying pattern, you'll be able to handle other patterns—as well as more intricate projects—with little difficulty.

Preparation

The simplest way to pour a footer for a brick wall is to use the trench as the form for the concrete. All the details on excavating and building footers can be found in chapter 7, but it's a good idea to consider the following suggestions:

- Choose a spot where the soil is firm and drainage is good. Avoid locating the wall near large trees; the roots can put excessive pressure on the foundation and cause cracking.
- Wall footers should be 2" to 3" below the ground, so that no concrete is visible around the wall.
- The footer should be twice the width of the wall and at least 6" thick. If it's being placed below the frost line, you might want to make it even thicker.
- Using 3/8"- or 1/2"-diameter rebar strengthens the footer considerably.
- Allow the footer at least two full days to cure before laying any brick.
- Stack your bricks in several piles along the job site. This will keep you from having to transport bricks in the middle of the job.
- Hose down all of the bricks a few hours before beginning the work. The hose will

also come in handy for keeping the mortar moist and rinsing your tools as you go along.

- With a tape measure, locate the outer edge of the wall by measuring in from the edge of the footer at each end. Snap a chalk line between the two points to mark a guideline to keep the wall centered on the footer.

Preparing the Mortar

1. Mix the mortar with water until you obtain a smooth, plastic-like consistency. Further details on working with mortar can be found in chapter 11.
2. Make a dry run by laying a course of stretcher bricks along the chalk line for the entire length of the wall. Leave 1/2" between each brick for the head joints and mark the position of the bricks on the foundation with chalk. Lay this course without cutting any of the bricks; adjust the head joint width if needed.

Making a dry run

Use the point of the trowel to furrow the mortar.

3. Remove the dry course from the foundation, then throw a mortar line on the foundation. To do this, load the trowel with mortar; as you bring your arm back toward your body, rotate the trowel to deposit the mortar evenly. It should be applied approximately 1" thick, one brick wide, and three to four bricks long. (You might want to practice throwing lines on the mortarboard until you become comfortable with the technique.)
4. Furrow the mortar with the point of the trowel using a stippling motion. Divide the mortar cleanly with the trowel; do not scrape it. Good furrows not only ensure that the bricks are laid evenly, but they also help to squeeze out excess mortar on the sides as the brick are set in place.

Laying the Brick

1. Lay the first course of stretcher bricks in the mortar. Beginning with the sec-

Checking wall for trueness

The two courses must be of equal height.

Checking that the bricks are level

ond brick, apply mortar to the head joint end of each brick, then shove the bricks firmly into place so that the mortar is squeezed out of all sides of the joints. Use a level to check the course for correct height, then place it on top to make sure that all the bricks are true and level.

2. Make sure that the head joint thicknesses correspond with your chalk marks. When you have to move a brick, tap it gently with the trowel handle; pull-

ing on it breaks the bond. Be sure to trim off any excess mortar from the sides of the bricks.

3. Throw another mortar line alongside the first course, then begin laying the other bottom, or backup, course. Use

The brick set is the best tool for cutting bricks.

the level to make sure that the two courses are of equal height, but do not mortar them together.

4. Cut two bricks to half length. To cut a brick, lay it on the ground and score it all the way around using a hammer and brick set. Break it in two with a sharp blow to the brick set. **Note:** When cutting bricks, protect your eyes by wearing goggles.

5. Use the two half bricks to begin the second, or header, course. This will ensure that the bottom two courses are staggered for structural purposes.

6. To finish the second course of the lead, lay three header bricks and make sure that they are true and level. (The lead is simply the end or corner of a wall.) The

Make sure that the header bricks are plumb and level.

third and fifth courses consist of stretchers, similar to the first course; the fourth course begins with a single header, followed by stretchers. Use the level on every course to make sure that the lead is true.

Using cut bricks allows you to stagger the courses.

The third and fifth courses are similar to the first.

The leads are ready to fill in.

A mason's line is helpful in filling in the leads.

Use a concave jointer to finish the joints.

7. Build another lead on the other end of the foundation. As the mortar begins to set, stop laying bricks and use a jointer to finish the mortar joints. Work along the vertical joints first; this will make them weatherproof as well as improve the appearance of the wall.

Filling in the Leads

1. Stretch a mason's line between the completed leads, then begin laying the outer course. The line should be approximately 1/16" away from the bricks and flush with their top edges. Work from both ends of the wall toward the middle.

Push the brick straight down to squeeze out the mortar from the joints.

When you reach the final brick, mortar both sides of it and push it straight down to squeeze out the mortar from the joints.
2. Move the mason's line to the back of the wall and begin laying the backup course. Remember to check your work with the level for accuracy and finish the joints with the jointer when almost dry.
3. The fifth, or top, course is laid exactly like the first. Move up the mason's line,

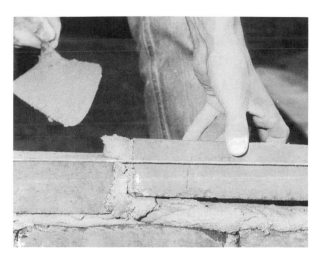

Laying the top course

throw a mortar line, and begin laying the bricks. Apply a generous amount of mortar to the head joint end of each brick, then shove it firmly into place.

4. For a higher wall, simply build more five-course leads at each end. Keep in mind that some type of reinforcing should be used for higher walls.

Applying mortar to the head joint end of the brick

Filling in the joints on the top course

The completed wall

5. Scoop mortar onto the trowel and use the jointer to fill in the joints on the top course. Keep a careful check on the joint thickness as you go. When the last brick has been laid, check the top course for alignment.

Building Corners

A wall with corners is not much harder to build than the basic freestanding wall. The following directions show how to build a corner in the common bond pattern, but they can be adapted to any of the other patterns as well.

1. Snap chalk lines on both sides, then check to make sure that they're perfect-

Laying the remaining bricks in the first course of the lead

Make sure the chalk lines are perfectly square.

The completed backup course

ly square using a carpenter's square or the 3-4-5 method.

2. Make a dry run to mark the position of the bricks. Throw a mortar line, then place the first brick exactly at the corner, being careful to line up with the chalk lines.
3. Lay the four remaining bricks in the first course of the lead. With a level and/or carpenter's square, check the alignment

Line up the first brick with the chalk lines.

and make sure that the bricks are level and plumb.

4. Throw mortar lines and lay the backup course. Both courses should be level with one another; there is no mortar joint between the two.
5. To lay the second course, cut two bricks into quarter and three-quarter pieces. Begin by laying the three-quarter pieces perpendicular to one another to form the outer edge of the corner. Continue by laying several header bricks out from the corner. Then complete the second course by inserting the two quarter bricks as closure pieces.

Completing the second course with closure pieces

6. Lay courses three through five to finish the corner lead. Courses three and five are similar to course one; course four begins with a header.
7. Construct a second lead at the opposite corner.

Checking the alignment of the corner lead

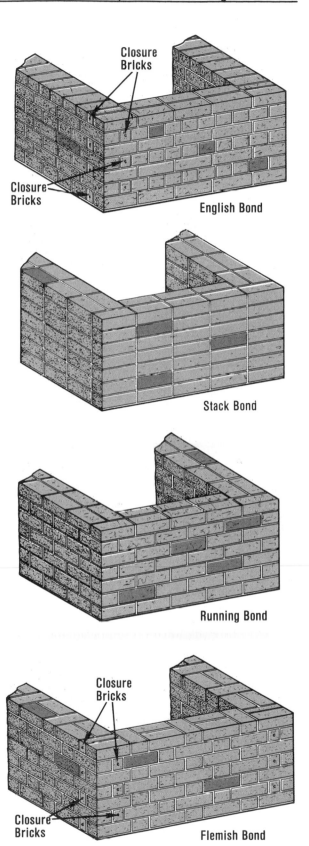

Other popular double wythe corner treatments

Closure Bricks

Closure Bricks

English Bond

Stack Bond

Running Bond

Closure Bricks

Closure Bricks

Flemish Bond

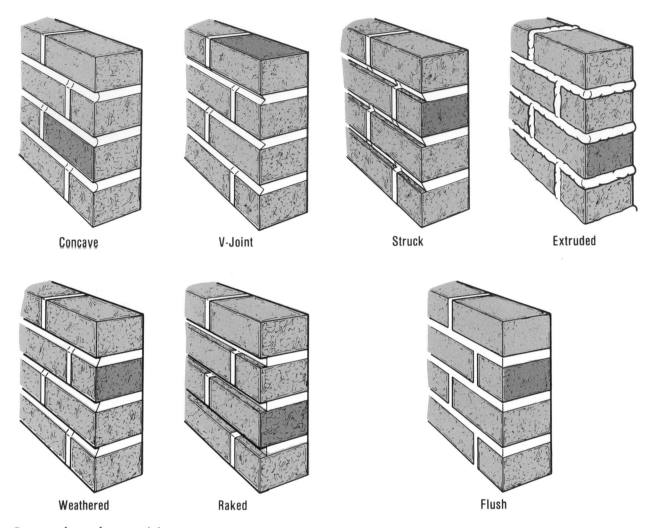

Concave   V-Joint   Struck   Extruded

Weathered   Raked   Flush

Commonly used mortar joints

MORTAR JOINTS

There are several methods of finishing mortar joints. Choose one that's based on the type of construction. The best joints for strength and waterproofing are concave and "V"-joints. A weathered joint is also strong and very watertight. Raked, struck, and extruded joints are perhaps the most attractive looking; however, they're not very water resistant. Care should be taken when using them in rainy or freezing climates. Struck joints are the least durable of the three, because they provide a ledge where dirt and moisture can accumulate. A flush joint is the simplest—excess mortar is simply cut off with the trowel. But this joint isn't particularly strong or water resistant.

SIMPLE BRICK PROJECTS

As pointed out earlier, brick is an extremely versatile building material. Following are some basic project ideas; with a little imagination, you're sure to come up with some of your own.

Pillars and Posts

The first step in making brick pillars and posts is to construct adequate footings. Once the

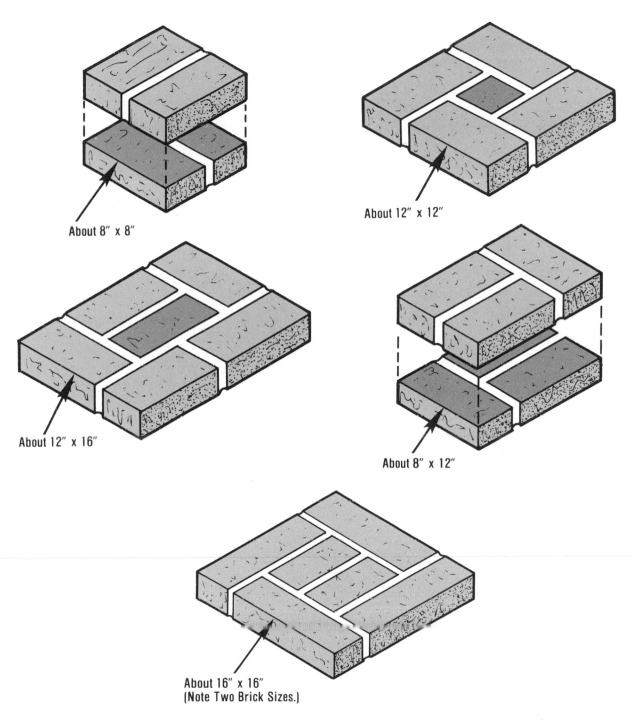

About 8" x 8"

About 12" x 12"

About 12" x 16"

About 8" x 12"

About 16" x 16"
(Note Two Brick Sizes.)

Bond patterns for solid posts and pillars; because the 8"×8" and 8"×12" styles are so narrow, alternate the courses for greater stability.

footings have set completely, build the pillar or post the same as if you were putting up a wall. Use a level to check the horizontal plane of each course and the plumbness of the whole struc-

ture. To make sure the corners are square, use a large carpenter's square.

When it comes time to top off the pillar or post, you have several options. A removable

Second. Fourth. Sixth. etc.. Courses

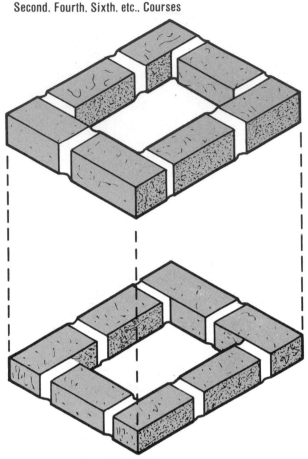

First. Third. Fifth. etc.. Courses

Bond pattern for a hollow post or pillar, with alternate courses

wooden cap or a piece of flagstone works well, or you can simply hand form a mound of concrete or mortar. For a more elegant look, use a precast concrete slab; details on slabs can be found in chapter 8. Whichever type of cap is used, install it at a slight angle so that it sheds water.

Brick and Redwood Bench

The normal seating height for this outdoor bench is 16" to 20", and the width from 20" to 24". As for length, 8' is good, although you can, of course, change any of the dimensions to suit your needs.

1. Stake out the perimeters of the two support walls, excavate to below the frost

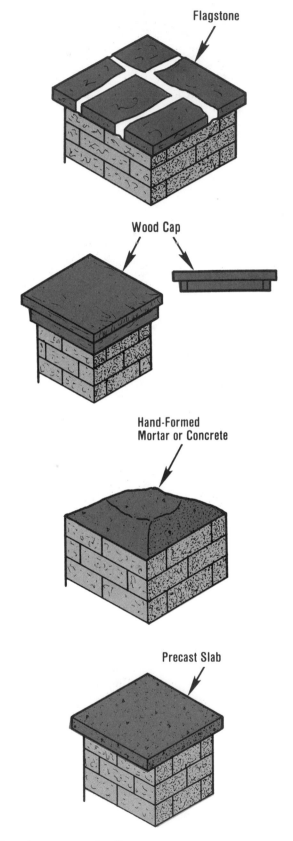

Caps for posts and pillars

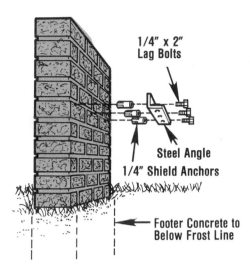

Angle iron is used to support the seat section.

Nailing the seat section together

line, and pour the footers to within a few inches of the grade.

2. Build up the support walls so that they're three or four courses higher than the seat level. Allow the mortar to cure for a few days.

3. Drill 1/4″-diameter holes in the walls and install the lag shield anchors. Attach the seat support angle iron by installing 2″-long lag bolts into the shield anchors.

4. Cut redwood 2×4s to length for the seat. You'll also need 1/4″×1″×3-1/2″ spacers.

5. Fasten the 2×4s and spacers together with galvanized nails, blunting the nails slightly to keep the wood from splitting. Locate spacers at the ends and midpoint of the seat.

6. Position the seat on the angle support. Drill pilot holes in the seat and fasten it to the support with 1/4″×2″ lag screws.

Planter

This attractive and functional planter sits on two 2-1/4″×12″×24″ precast concrete slabs. Set the slabs on a 2″ layer of sand below grade. The bottom course of bricks is laid dry; the vertical joints will act as weep holes for drainage. The remaining courses are laid up in the standard fashion.

For a decorative effect, you can leave one or more of the top courses overhanging slightly. This process is known as *corbelling*. As a final step, apply a 3/4″-thick layer of mortar to the

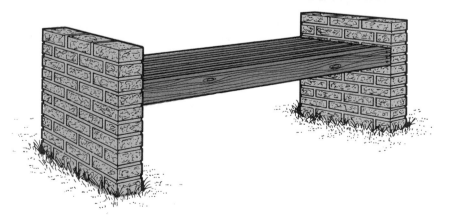

The completed bench

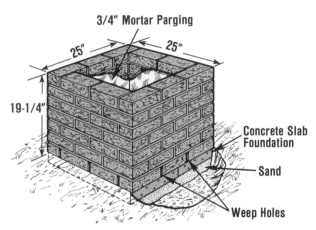

A brick planter is a handsome addition to your home's exterior.

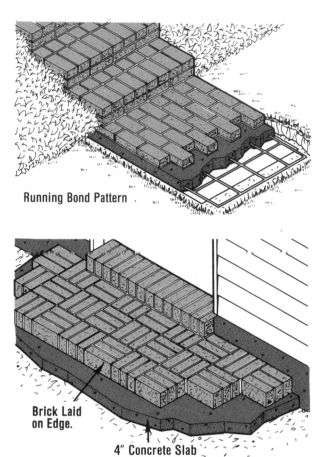

inside of the planter. By bringing this parge coat up to the top course, the waterproofing qualities of the planter will be improved.

Brick-Faced Steps

As an alternative to plain concrete steps, bricks can be used as facing over the concrete. Building brick steps involves the same procedures that are used for concrete steps to prepare the base, build the forms, pour the concrete, finish the surface, and cure the concrete. Details on concrete steps can be found in chapter 7. To lay the bricks in the concrete, use the following procedure:

1. Wet down the bricks a few hours before beginning the project. This will prevent them from absorbing too much water from the mortar.
2. Place and screed a 1/2"-thick wet mortar bed between temporary form boards set one brick length apart. Use a special bladed screed to level the concrete. The screed rides on the forms and extends down one brick thickness below them. The forms should be set for this depth, plus an extra 1/2" to allow for the mortar bed.
3. As the screed is moved along, place the bricks in the wet concrete, leaving 1/2"

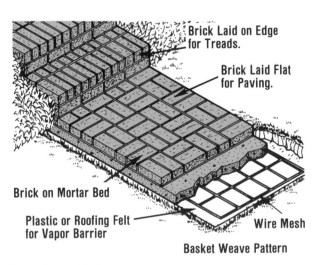

Variations of brick steps

open joints between them. A 1/2" wooden spacer can be used to ensure that the joints are uniform. Gently tap the bricks with a wooden mallet to set them in place.

4. Use a trowel to pack the concrete into the joints, then tool the joints with a concave jointer.

5. Wait several hours, then scrub the set bricks with a burlap sack to remove any stains. Once the mortar is dry, brush the surface with a stiff broom or bristle brush to remove any bits of dry mortar.

The bricks can be laid in the pattern of your choice, including the traditional running bond pattern shown in this example. Another popular choice is the basket weave pattern, with the bricks either laid flat or on edge.

Raised Tree Wells

Physically, a tree well is nothing more than a retaining wall. When the ground is cut away, a raised well keeps the soil surrounding the tree at its original level. This helps the tree's roots from becoming exposed and threatening the life of the tree.

1. After determining the well size, lay a length of rope around the tree where the interior edge of the well will be. Slide a piece of plywood under the rope and mark the arc of the circle on the wood.

2. The arc will mark the front edge of the template. Cut a 2'- to 3'-long template for the well out of the plywood.

3. Dig a 16"-wide trench, beginning about 4" inside the front edge of the template, several inches below the frost line. Keep the bottom 6" of the trench as smooth as possible so that a footer can be poured without the need for forms.

4. Pour the footer, screed it smooth, and level it. Cure for at least one full day.

5. Dry-lay two wythes of brick around the footer, keeping 1/2" joints between the wythes and bricks. The wythes are laid in a horizontal running bond pattern so

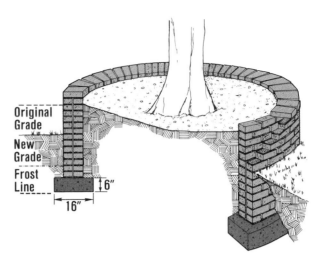

This raised tree well is both attractive and effective.

that the head joints don't extend through the width of the wall. Mark the location of the bricks on the footer.

6. Remove the bricks and lay a 1/2" bed of mortar on the footer.

7. Lay the first course in the mortar bed. Don't fill in the head joints or the joints between the wythes. These are left open so moisture can escape.

8. Plumb and level each brick, using the template to keep the well plumb from point to point.

9. Using 1/2" bed joints and a vertical running bond in each wythe, build up the well to the former ground level. Plumb and level as you go.

10. After each course is laid, clean out any mortar that falls into the head joints. Metal ties can be used to bond the wythes together every five to six courses.

11. Tool the joints with a jointer as they set up. After the joints harden, brush out any loose mortar.

12. Lay the top course with headers. Apply mortar to both sides of the bricks before laying them. Tool the top joints flat to prevent moisture entry.

Chapter 13

ADVANCED BRICK PROJECTS

If you've mastered the brick projects in the previous chapter and you're ready for more, this chapter's for you. Curved walls, fireplaces, and barbecues are among the projects discussed on the following pages. Keep in mind that none of these projects have to be followed to the letter; you can simply use them as the starting point for your own ideas and designs.

BRICK VENEERING

Veneer masonry is a popular choice for home building and remodeling. It gives the appearance of a solid brick wall while providing better economy and insulation. It can be used as an addition to conventional wood frame structures, and can also be placed on concrete block walls. You might want to cover an entire wall from foundation to roof, or you can stop at windowsill level.

Preparing the Foundation

If the footing of the house extends out 6" or more, the veneer can rest directly on top of it. If the footing is less than 6", it must be "extended" before the veneer can be placed on it. To do this, either bolt a corrosion-resistant steel angle to the existing foundation, or pour a new footer next to the foundation. If making a new footer, pour it to within a few inches of ground level, then start the veneer at that point. The veneer must always

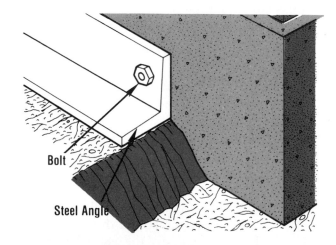

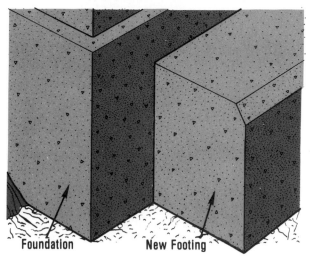

Two methods of extending a footer

be tied to the old foundation; to ensure a good bond, wash the old foundation surface, and coat it with grout.

196

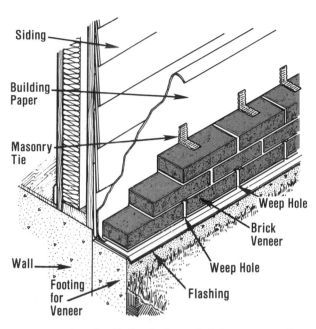

Construction details for laying a brick veneer wall

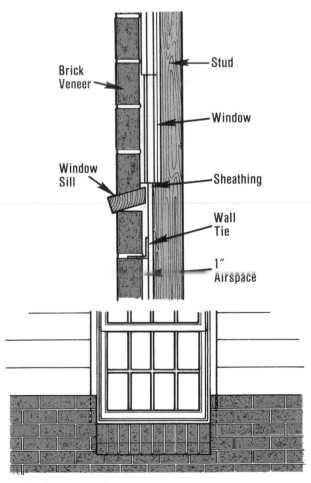

Windowsills require special treatment

Laying the Wall

1. Install flashing over the footer to prevent water from seeping behind the veneer. Use copper, aluminum, or roofing paper for this. Spread a 1/2" bed of mortar on top of the footer, then push the flashing down firmly into it.
2. Cover the existing siding with a good sheathing material, such as tar paper. Always leave a 1" airspace between the sheathing and the veneer.
3. To lay the first course of bricks, spread a 1"-thick bed of mortar on top of the flashing. Tap each brick into place with the trowel handle; never pull on a brick, because this can break the bond. Make sure that all the bricks are plumb and level.
4. To hold the veneer in place, nail galvanized metal wall ties through the siding and into the studs. Space the ties every 32" horizontally and every 16" vertically, and offset the rows so that the ties don't line up.
5. Make weep holes, approximately 24" on center, in the vertical joints of the first course of bricks. To form the holes, use short lengths of rubber tubing or cord, which can be easily removed after the mortar has set. Weep holes allow any water that might seep in to escape despite any flashing.
6. Continue laying each course of bricks, being careful to maintain the precise mortar joint thickness needed for the desired wall height. A story pole can be helpful for greater accuracy. In addition, take care to maintain the 1" airspace the entire height of the wall.
7. When you reach a window sill, lay the bricks on edge in rowlock fashion. The bricks should also be installed on a slant in the direction of the rainfall.
8. For bricks being laid above windows and doors, a steel lintel must be used as a base. The lintel is set onto the course, even with the head of the window or

197

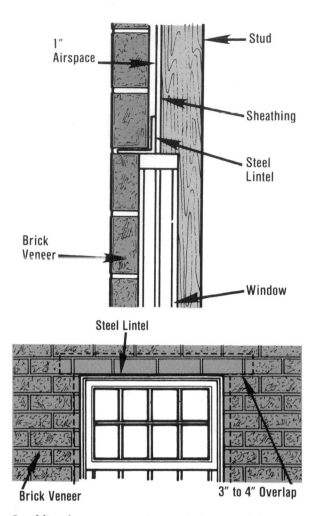

Steel lintels are a must above windows and doors.

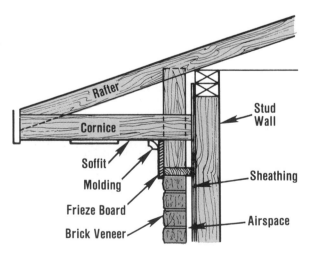

The veneer must meet the frieze board.

door. It must overlap the bricks by 3" to 4" on both sides of the window or door.

9. If the veneer is being carried all the way to the roof, it must meet the frieze board. The frieze board on the cornice should overlap the top course of bricks by at least 1/2". Use 2×4 blocking to provide a sound nailing surface for the frieze board.

SERPENTINE PLANTER WALL

Although this project might look difficult at first glance, it really isn't; the most challenging part of making the planter wall is the template. Start with a 4'×8' sheet of 1/2" plywood and cut two pieces from it: one 24"×48" and the other 24"×96". The rest of the procedure is as follows:

1. Use cleats to nail the two pieces of plywood together. You now have a 24"×144" panel from which to make the template.

2. Lay the panel on a flat surface and establish two base lines at right angles to one another. Draw one of the lines 2" in from one long edge of the panel and the other along a short edge. Use a carpenter's square to make sure the lines are perpendicular.

3. Measuring from the base lines, mark the location of the radius centers with a pencil.

4. Use a straightedge and pencil to draw the radii on the template. When the pattern is complete, cut out the area between the lines.

5. Stake the two pieces of the template in place on the ground, making sure you maintain an 8" opening the entire length. The easiest way to do this is to maintain 24" between the outside edges of the pieces.

6. Dig a 12"- to 18"-deep trench along the curve of the template. Fill the trench

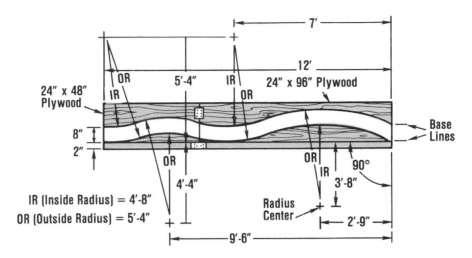

Serpentine planter wall template

Digging the trench

with concrete to within a couple of inches of the grade. Allow it to cure for a few days.

7. Use the template as a guide for laying the bricks. First make a dry run to see how well one course of full bricks fits and to adjust the width of the vertical joints between them. Stagger the end joints of the two rows of bricks.

8. Lay the first course of bricks. Place metal ties across the rows at 24" intervals to tie them together. Do this every couple of courses.

9. As you lay the course just above final grade, make weep holes through the wall approximately 12" to 18" apart.

10. Check the wall as you go to make sure that it's plumb and level. This is necessary so that it will conform to the shape of the template as you lay additional courses.

11. Cap the wall with a rowlock, as shown. For a more finished-looking corner, miter-cut the bricks.

12. Allow the mortar to cure for a few days, then coat the foundation wall with wa-

199

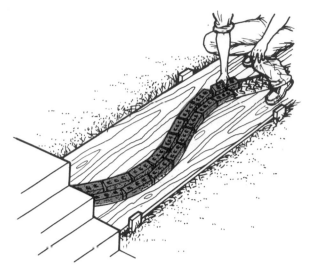

Make a dry run to check the fit of the bricks.

The completed planter wall

Double Stretcher Garden Wall Bond

Triple Stretcher Garden Wall Bond

Examples of advanced bond patterns

terproofing material. Line the bottom of the planter with 15-pound building felt or plastic sheeting to provide a moisture barrier.

13. Place 6″ of sand or gravel for drainage. Cover it with a generous layer of topsoil, and you're ready to add the plants and flowers.

ADVANCED BONDING PATTERNS

The bond patterns discussed in chapter 12 can be enhanced for more visually striking results. For instance, the look of Flemish bond can be greatly altered simply by increasing the number of stretchers between the headers in every course. Two stretchers between each header are known as a *double stretcher garden wall bond;* three stretchers between each header are called a *triple stretcher garden wall bond.* The completed bond forms a repeating diamond pattern in the brick wall; using different colored bricks for the headers creates an added effect.

CREATING OPENINGS IN BRICKWORK

As you progress in laying a brick wall, you might have to contend with openings in the brickwork. Windows, doors, and circular openings are all commonplace, so it's important to be able to deal with them safely and effectively.

Windows and Doors

All window sills should rest on the same course. When the distance from the foundation

A lintel provides additional support to a brick wall.

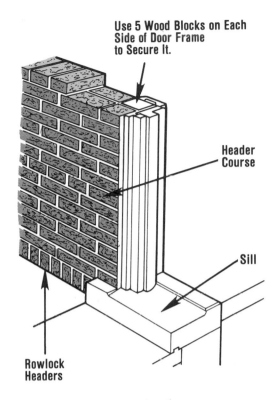

Use 5 Wood Blocks on Each Side of Door Frame to Secure It.

Header Course

Sill

Rowlock Headers

Door opening construction details

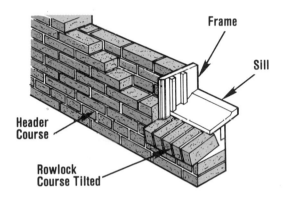

Frame

Sill

Header Course

Rowlock Course Tilted

Window opening construction details.

to the bottom of the window sill is known, the number of courses required to bring the wall up to sill height can be determined. If the sill is to be 4'–4-1/4" above the foundation and 1/2" mortar joints are being used, 19 courses will be required. Each brick plus a mortar joint is 2-1/4" + 1/2" = 2-3/4". One course is thus 2-3/4" high. Four feet, 4-1/4" divided by 2-3/4" is 19, the number of courses required.

With the bricks laid to sill height, the rowlock sill course is laid. This course is pitched downward normally, taking up a vertical space equal to two regular courses of brick. The exterior surface of the joints between the bricks in the rowlock course must be carefully finished to make it watertight.

The window frame is placed on the rowlock sill as soon as the mortar has set and is temporari-

ly braced until the brickwork has been laid to about one-third the height of the frame. These braces should be kept in place for several days so that the wall above the window frame will set properly.

The remainder of the wall is laid so that the top of the bricks in the course level with the top of the window frame is not more than 1/4" above the frame. Wall height can be adjusted by adding or removing courses, expanding or reducing the joint, or a combination of both. The corner leads should be laid after the height of each course at the window is determined.

Lintels are placed above windows and doors to support the weight of the wall above them. They rest on the brick course that's level or approximately level with the frame head and are firmly bedded in mortar at the sides. Lintels are made of steel, precast reinforced concrete beams, or wood. The recommended thickness of the angle for a two-angle lintel is 1/4". This makes it possible for the two-angle legs that project into the brick to fit exactly in the 1/2" joint

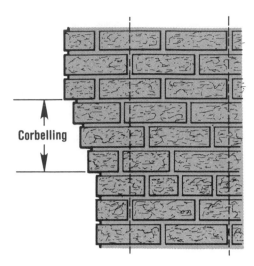

A corbelled brick wall

between the face and backup ties of an 8″ wall. The space between the window frame and lintel is closed with blocking and weatherstripping. The wall is then continued above the window after the lintel is placed.

The same procedure used for laying brick around a window opening, including placement of the lintel, can be used for laying brick around a door opening. The frames of doors and windows are shimmed with wedges to square them. The expansion anchors or lead shields are installed next to provide a means for securing the frames.

Corbelling is a procedure in which courses of brick are set out beyond the face of the wall to form a self-supporting projection. The exposed portion of a chimney is frequently corbelled out and increased in thickness to improve its weather resistance. Headers should be used as much as possible in corbelling. The first projecting course may be a stretcher course, if necessary, and no course should extend out more than 2″ beyond the course below it. The total projection of the corbelling should not be more than the thickness of the wall. Corbelling must be done carefully, and all mortar joints should be completely filled.

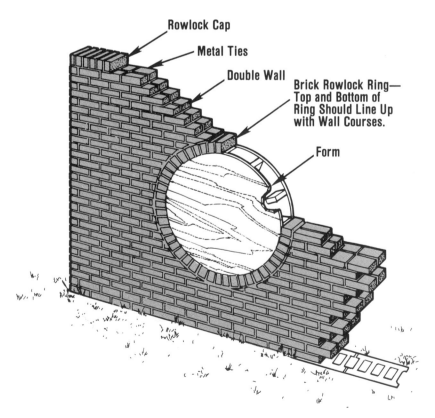

The form is encircled with a rowlock made of full bricks.

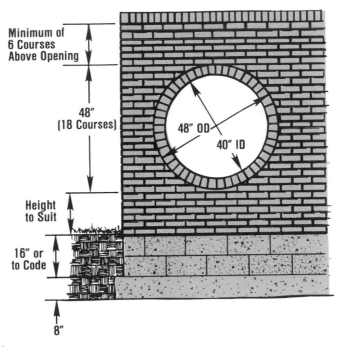

Minimum of
6 Courses
Above Opening

48"
(18 Courses)

48" OD
40" ID

Height
to Suit

16" or
to Code

8"

Elevation view of the wall

Circular Openings

A round opening in a garden or patio wall is a real eye-catcher. In the example shown, the wall is laid in the running bond pattern, with no headers used. Metal wall ties are needed to tie the wall together; use them every six courses for maximum strength.

The key to this project is the circular form. It's built with two pieces of 1/2" plywood, each one 40" in diameter to match the inside diameter of the wall opening. Cross bracing made from 6-1/2" long 2×4s is nailed to the inside of the plywood to make the form 8" wide so that it can carry the brick rowlock ring. The rowlock spaced around the opening is marked on the outside edge of the form using a modular spacing rule. Be sure the rowlock is made of only full bricks; don't use any partials.

Build up the wall to the point where the opening starts, then set the form in place and brace it. Build in the arch bricks on the form as you continue the wall. It will take eleven courses to complete the opening; the last brick laid should be even with the top of the arch. A minimum of six courses should be built over the top

of the opening. Make the last course a solid brick cap to prevent water from entering.

Wait at least two days before removing the form. Neatly trim out any hard mortar in the joints with a chisel, and repoint with fresh mortar. Don't worry about the soundness of the structure—the circular opening can support the bricks laid over it even better than a lintel.

FIREPLACES AND CHIMNEYS

Your local building codes will likely specify fireplace and chimney requirements. For a chimney that's multistory, extra wide, or extra high, special design considerations are needed that should be left to the professional. Although a fireplace is expected to be an attractive architectural feature, its basic function is threefold: (1) to assure proper fuel combustion, (2) to deliver smoke up the chimney, and (3) to give off the maximum amount of heat possible.

Proper fireplace operation depends on the shape and dimensions of the combustion chamber; the location of the fireplace throat and

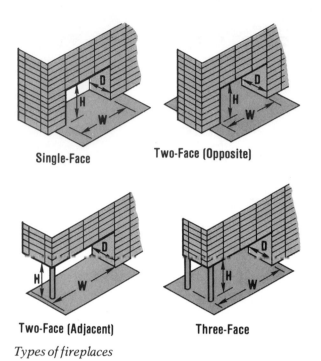

Single-Face	**Two-Face (Opposite)**
Two-Face (Adjacent)	**Three-Face**

Types of fireplaces

Fireplace Elements

The traditional single-face fireplace is most common, and the majority of design information is based on this type. Other types are explained in table 13-1. Following are the major elements of a fireplace; you should understand each of them and their function.

- Hearth. The floor of the fireplace is called the hearth. The inner part of the hearth is lined with firebrick; the outer hearth consists of noncombustible material such as firebrick, concrete brick, concrete block, or concrete. The outer hearth is supported on concrete.
- Lintel. The lintel is the horizontal member that supports the front face of the fireplace above the opening. It can be made of reinforced masonry or a steel angle.
- Firebox. The combustion chamber where the fire is made is called the *firebox*. Its side walls are slanted slightly to radiate heat, and its rear wall is curved or inclined to provide an upward draft to the throat.

smoke shelf, and the size of the flue area in relation to the fireplace opening. Fire safety depends on a good design and on the ability of the bricks or other masonry units to withstand high temperatures.

TABLE 13-1: FIREPLACE TYPES AND STANDARD SIZES					
Type	**Width (w) (in.)**	**Height (h) (in.)**	**Depth (d) (in.)**	**Area of Fireplace Opening (sq. in.)**	**Nominal Flue Sizes (Based on 1/10 Area of Fireplace Opening) (in.)**
Single-face	36	26	20	936	12×16
	40	28	22	1,120	12×16
	48	32	25	1,536	16×16
	60	32	25	1,920	16×20
Two-face (adjacent)	39	27	23	1,223	12×16
	46	27	23	1,388	16×16
	52	30	27	1,884	16×20
	64	30	27	2,085	16×20
Two-face (opposite)	32	21	30	1,344	16×16
	35	21	30	1,470	16×16
	42	21	30	1,764	16×20
	48	21	34	2,016	16×20
Three-face	39	21	30	1,638	16×16
	46	21	30	1,932	16×20
	52	21	34	2,184	20×20

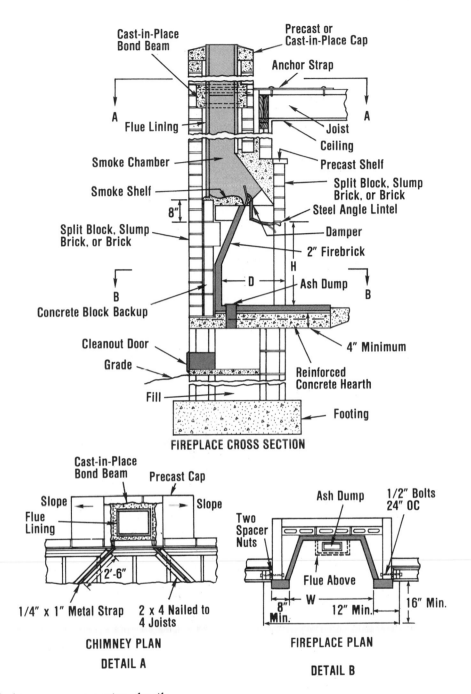

FIREPLACE CROSS SECTION

CHIMNEY PLAN
DETAIL A

FIREPLACE PLAN
DETAIL B

Fireplace and chimney construction details

Unless the firebox is the preformed metal type, it should be lined with firebrick that is at least 2" thick and laid with thin joints of fireclay mortar. The back and side walls, including the lining, should be at least 8" thick to support the chimney weight.

The fireplace is laid out on a concrete slab and its back is built up about 5' high before the firebox is constructed and backfilled with tempered mortar and brick scraps. Backfilling solidly behind the firebox wall will allow for some expansion of the firebox.

- Throat. The throat of a fireplace is the slot-like opening directly above the firebox through which flames, smoke, and other combustion products pass into the smoke chamber. The throat must be carefully designed to permit proper draft. It must be at least 6″ above the highest point of the fireplace opening.

  The inclined back of the firebox extends to the same height as the throat and supports the hinge for the metal damper. The damper extends the full width of the fireplace opening and should open upward and backward.

- Smoke Chamber. The smoke chamber compresses and funnels smoke and gases from the fire into the chimney flue. The shape of this chamber should be symmetrical with the centerline of the firebox to assure even burning. The back of the smoke chamber is usually vertical, and its walls are inclined upward to meet the bottom of the chimney flue lining. If the solid masonry wall is less than 8″ thick, the smoke chamber should be coated with 3/4″ of fireclay mortar. Metal lining plates are available to give the chamber its proper form, provide smooth surfaces, and simplify construction.

Chimney Elements

A fireplace or wood-burning stove chimney creates a draft and disposes of combustion products. To prevent downward air currents, build the chimney at least 3′ above a flat roof, 2′ above the ridge of a pitched roof, or 2′ above any part of the roof within a 10′ radius of the chimney. Increasing the chimney height will improve the draft. Here are a couple of other critical factors:

- Foundation. The concrete foundation for a chimney is designed to support the weight of the chimney. The footer thickness should not be less than one and one-half times the footer projection. The bottom of the footer should extend below the frost line.

- Chimney Flue. A flue must have the correct dimensions and shape to produce the proper draft. Smoke should be drawn up the flue at a relatively high velocity. The total area of the flue opening should be approximately one-tenth of the fireplace opening area. Check your local codes.

A fireplace or stove chimney may have multiple flues, but each one must be built as a separate unit. The American Insurance Association requires clay flue liners for all residential fireplace chimneys. Flue liners extend from the top of the throat to at least 2″ above the chimney cap. Chimney walls are constructed around the liner segments. The space between all masonry joints should be completely filled with mortar.

Minimum wall thickness, measured from the outside of the flue liner, should be 4″. Exposed joints inside the flue should be struck smooth. When a chimney contains more than two flues, they should be separated by 4″-thick masonry bonded into the chimney wall. The tops of the flues should have a height difference of 2″ to 12″ to prevent smoke from pouring from one flue into another.

Chimneys should be built as vertical as possible, but a slope is allowed if the full area of the flue is maintained throughout its length. Slope should not exceed 7″ to the foot or 30 degrees. Where offsets or bends are necessary, miter both ends of adjoining flue liner sections equally to prevent reduction of the flue area.

OUTDOOR BARBECUE

To provide a solid base for the barbecue shown, build it in a corner of your patio. To begin, measure and stake off the desired area. The perimeter for the barbecue pit should be approximately 3′×3′. The walls of the excavation can be used as forms for the concrete, provided the soil is firm enough.

Set a 22″-diameter cardboard form tube into the center of the pit, using 2×4s to keep it from moving. The form tube must be long enough to extend from the bottom of the excavation to a point 26-1/2″ above ground. Make sure that it's

Make sure the form tube is plumb and level before filling it with sand.

plumb and level, then fill the tube with sand to approximately 2″ below the final grade. Pour concrete mix around the base of the tube, completely filling the trenches. Screed the footer and allow at least three days for the concrete to cure.

Laying the Inside of the Barbecue

1. Begin laying half-bricks, with the cut ends facing out, around the form tube. Use 1/2″ mortar joints. Be sure to leave an opening for the ash door.

2. When the third course has been completed, install a 3/8″×2″ horizontal steel bar across the opening. This will support the bricks above the opening.

3. When the seventh course has been completed, drill three equally spaced 3/8″-diameter holes in the tube. Insert 3/8″-diameter bent rods to support the metal grate, then lay two additional courses.

4. Drill three more 3/8″-diameter holes, insert three more rods to act as supports for the grill, then lay the final course of half-bricks. The inside of the barbecue is now complete.

Laying the Outside of the Barbecue

1. Lay the first wythe of the rectangular outside wall in a running bond pattern; build it to a height of four courses. Build the corners first and work toward the center, using a level to check for plumb and level and to keep the wall aligned.

2. Construct temporary 2×4 supports and set them against the brick, flush with the fourth course.

3. Lay a course of bull headers.

4. Build two wythes of stretchers in a running bond pattern two courses high.

5. Lay a course of bull headers level with the height of the pit wall.

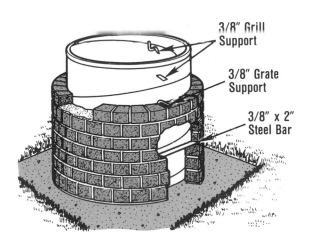

Building the interior of the barbecue

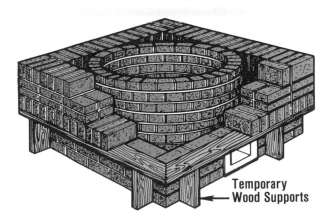

The outside of the barbecue is made up of four courses.

207

Filling between bricks with concrete

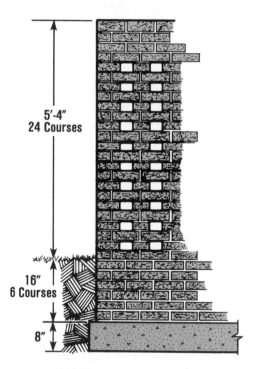

5'-4"
24 Courses

16"
6 Courses

8"

All courses below the grade line must be solid brick or block.

6. After the mortar joints begin to set, finish the joints and brush off any loose mortar particles. The top joints should be struck flat to prevent moisture entry. Remove the supports after twenty-four hours.

7. Peel out the form tube down to the sand. Smooth the sand and lay in a 4" concrete bed, sloping slightly toward the ash opening for water drainage.
8. Fill the spaces between the interior and exterior wall with concrete, flush with the brick height. Install the grate and grill. Cure the concrete a full ten days before using the barbecue.

BRICK SCREEN WALL

The type of brickwork used to create this screen wall is known as *latticework*. It takes more time to build than a regular brick wall, and the half-bricks must line up vertically. Once the excavation is finished and the footer has been poured, note that all of the courses to just above the grade line consist of solid brick; you can substitute 4" concrete block if you want, since it works equally well. The procedure for building the screen portion is as follows:

1. To start the first screen course, lay a 6" cut brick in mortar at each end of the wall. Attach a chalk line between the two bricks.
2. Lay a cut half-brick in place by centering it on the joint where the two bricks meet on the course directly below it. The cut part of the brick won't show, since it's laid with the cut end facing the back of the wall.
3. Leave a 4" gap, then lay another half-brick. The rest of the course is alternating gaps and half-bricks.
4. Start the second course with a half return brick on the corner. Follow with a whole brick extending over to the middle of the half-brick on the course below. The rest of this course is made up of whole bricks. The third screen course repeats the first, the fourth course repeats the second, and so on up the wall.
5. Compress the mortar joints fully around all the gaps, especially on the sides, to better stabilize the bricks.

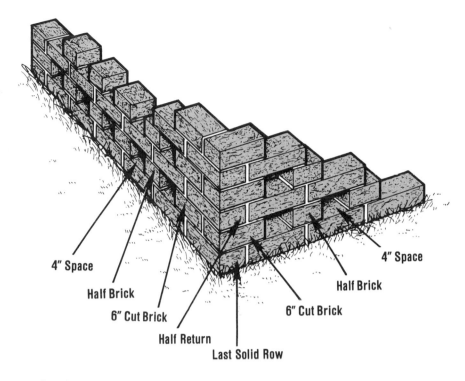

4" Space

Half Brick

6" Cut Brick

Half Return

Last Solid Row

4" Space

Half Brick

6" Cut Brick

Screen wall corner detail

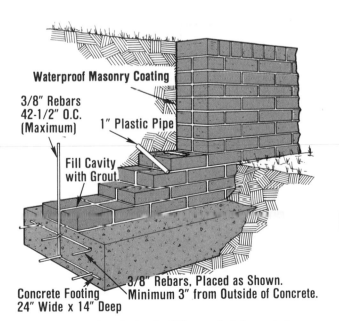

Waterproof Masonry Coating

3/8" Rebars 42-1/2" O.C. (Maximum)

1" Plastic Pipe

Fill Cavity with Grout.

3/8" Rebars, Placed as Shown. Minimum 3" from Outside of Concrete.

Concrete Footing 24" Wide x 14" Deep

Construction details for building a brick retaining wall

6. Check the vertical alignment of the halves with a level at least every three courses to make sure that they're not shifting.

7. Until the mortar fully sets, a screen wall is vertically weak. As you continue building the screen, be careful not to put too much pressure on the wall. It could be pushed over easily, or at the very least the mortar joints could be broken.

8. For decorative effect, most people prefer finishing off the wall with three or four solid brick courses.

BRICK RETAINING WALL

As with the concrete retaining wall discussed in chapter 7, this brick version provides valuable and long-lasting protection from soil erosion. Don't forget to check your local building codes and ordinances before beginning work on any retaining wall. Many municipalities require a building permit and an engineer's approval before a wall higher than 3' can be built.

1. For added structural integrity, insert steel rebar between the wythes. Pour a

grout mixture between the wythes to bond the rebar in place.

2. Apply a coat of waterproof masonry coating or sealer to the back of the wall to make it watertight.

3. Lateral or through-wall drainage must be provided to prevent water pressure from building up behind the wall. Details are the same as those illustrated for concrete retaining walls in chapter 7. Simply notch the brick to accommodate the through-wall pipe as you build the wall. Seal around the pipe with mortar.

4. Backfill around the lateral drainpipes or through-wall drains using crushed stone or gravel, not tight-packing soils that could cause clogging.

OTHER PROJECT IDEAS

From simple lawn edging to ornate entry ways, brick lends a distinctive touch to your property and home.

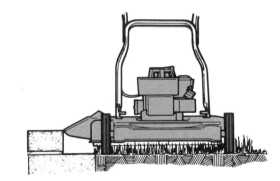

A

B

C

D

Brick project ideas: (A) garden edging, (B) mailbox post, (C) gate entrance, and (D) signpost treatments

Chapter 14

PAVING WITH BRICK AND CONCRETE PAVERS

Whether the project is a short walkway, a curving drive, or an expansive backyard patio or plaza, brick is a surprisingly easy material to work with. Although more expensive, precast concrete pavers are also popular because they're available in a variety of styles and colors. With these, you don't have to worry about getting a good finish, and you can work at your own speed—two obvious advantages over working with concrete.

Depending on the particular project design, most tools and materials needed can be found in the home. Whatever the project, proper drainage requires a slope of not less than 1" per 10 lineal feet. The maximum slope for a driveway is 1-3/4" per foot. Always slope away from buildings and toward pavement edges.

PATTERNS FOR BRICK

One of the advantages of paving with brick is that a pattern can be created in the ground. Brick lends itself to countless different patterns, from traditional to your own unique creations. Naturally, the easiest and least wasteful ones are those that don't require any of the brick to be cut.

There are three primary paving bonds: the herringbone, the running bond, and the basket weave. However, there are many combinations

Herringbone

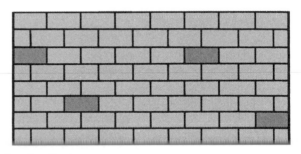

Running Bond

Basket Weave

The three primary paving bond patterns

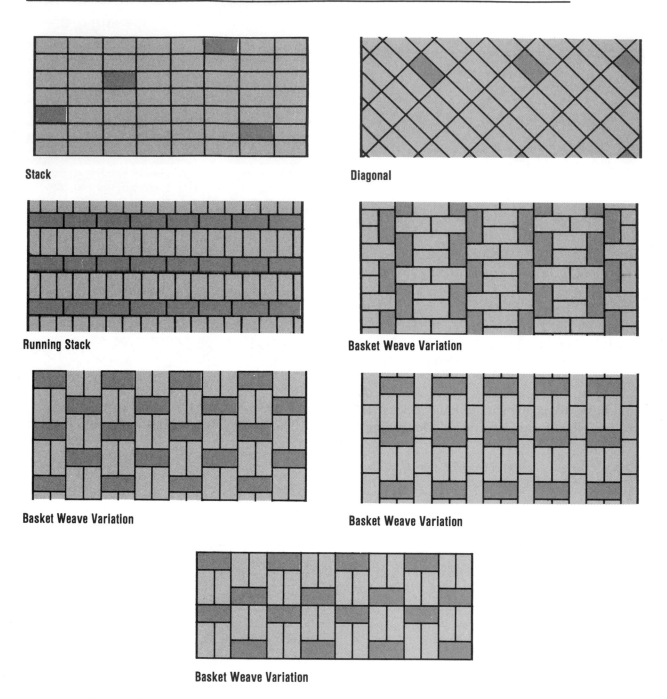

Stack

Diagonal

Running Stack

Basket Weave Variation

Basket Weave Variation

Basket Weave Variation

Basket Weave Variation

Variations of paving bonds

or variations of these. The running bond is the same pattern used in a brick wall. The herringbone is built of brick laid on a 45° angle from the border, with some cutting required along the edges. This pattern will resist movement better than any of the others, although it's more difficult to do. The basket weave is simply an alterna-

tion of stretcher bricks, much like a weave of yarn or cloth. All bricks vary slightly in length from burning in the kiln, so slight adjustments will have to be made as the work progresses. Paving bricks can be laid in a mortar bed or in sand, depending on your preference. Some general guidelines for paving follow.

Cutting

Brick, stone, and adobe are cut similarly. A brick set or broad-bladed cold chisel, mason's hammer, and safety glasses are necessary. The brick is set on a flat surface, and the chisel is placed on the cutline with the bevel facing away from the side to be used. A small groove is cut across all four sides with light hammer taps; a sharp blow on the broad surface will make the final cut. For larger stones, one end is placed on a solid support, and the unsupported end is tapped to make the final cut. Use the chisel end of the hammer to chip away any rough edges.

Laying the Pavement

Whatever pattern or material is used, laying begins at a corner and moves outward. Leveling is done in two directions. Except for brick on a sand base, all paving should have at least 1/2" joints to allow for the mortar fill. Depending on the look desired, the joints should be finished with a trowel or similar object after being filled. To prevent displacing previously laid bricks, a large piece of 1/2" plywood can be knelt on while laying the pavement.

Cleaning

Again, depending on the desired effect, you might want to clean the brick or stone if mortar stains appear. To do this after the mortar has set, soak the area to be cleaned, then mix a solution of twelve parts water to one part muriatic acid.

Kneel on plywood to distribute your weight evenly on the bricks.

Use rubber gloves to scrub the area, then rinse thoroughly to neutralize the acid.

COMPARING PAVING METHODS

Brick and other masonry units can be laid one of three ways: in sand, in dry mortar, or in wet mortar. Each technique has its advantages and limitations, so choose the one you think you'd be most comfortable with.

Laying Masonry Units in Sand

This is a simple method that yields surprisingly sturdy results. Once the surface is graded, a strong edge is built and a sand bed is then poured in and leveled. The units are placed tightly together, filling the area. Half-inch joints may be left in between, but this won't be as strong.

As the last step, sand is swept into the narrow joints between units, where it acts like a wedge to lock them together. Occasional resanding improves the bond strength. If the edges hold and the ground doesn't move, this paving is simple and permanent. Some settling over time is common, so build a little high to allow for it. Weeds in the joints can be kept down with a contact weed killer, or you can lay plastic sheet-

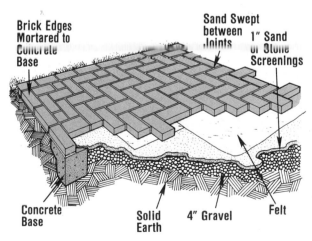

Cross-sectional view of masonry units with a sand base

ing or black felt tar paper between the bricks and the sand.

Paving without mortar is easy and relatively inexpensive. It provides a flexible surface that allows for easy repair should tree roots or freezing weather cause it to buckle. There are no mortar joints to crack from constant expansion and contraction. Also, if a unit is damaged, it can be easily replaced.

Dry or Wet Mortared Paving

Added stability is gained when you work with mortar, either wet or dry. Working with dry mortar is almost as easy as using sand, and it increases the permanence of the paving in areas subject to frost heave. The masonry units are bonded at the joints, providing good stability for driveways and the like.

In the wet mortar method, the masonry units are set in wet mortar over a 4″ to 6″ concrete slab. This permanent installation method is needed in areas with severe winters or highly unstable soil. If an existing slab is used as the base, repair all cracks and damaged areas first. If the existing slab is painted or extremely smooth, coat it with concrete bonding adhesive prior to

laying down the mortar bed. This will ensure a strong, lasting bond between the masonry units and slab.

EDGINGS

Strong edgings are the key to a secure pavement or walkway. The brick-in-soil type is the easiest to make; however, the ground must be very firm and capable of holding the bricks securely. Keep in mind that edgings should be installed after all excavation is finished, but before the paving begins.

There are two types of brick edging. One features the bricks standing side by side, with the full length of the bricks buried to prevent tipping. The other has the edging bricks tilted 45°. This not only produces a decorative notched effect, but it also allows the pavement to rise above grade while keeping as much of the edging as possible underground.

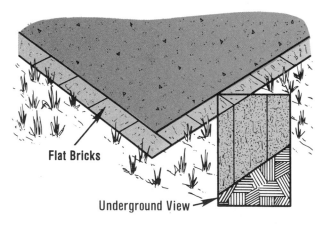

Flat Bricks

Underground View

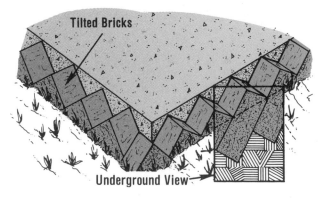

Tilted Bricks

Underground View

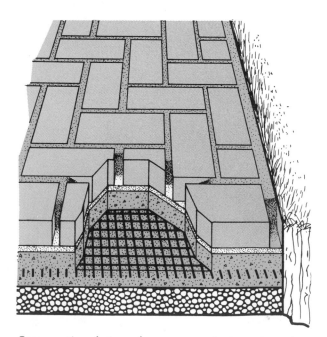

Cross-sectional view of masonry units in wet mortar

Two types of brick edging

To install either type of edging, dig a narrow trench just wide enough to fit the bricks. Use a level to make sure the edging is of uniform height. Pack the earth tightly against the outer perimeter of the bricks to secure them.

PAVING WITH A SAND BASE

For mortarless paving using any pattern other than running bond, you should use bricks whose length is twice their width, such as 4×8 units. This will make for a more sturdy and permanent construction. Because mortarless paving has a tendency to spread or creep slightly at the edges under normal traffic conditions, edging should always be used. You can use bricks, or 2″ wood framing also works well. Edging can also be used as a guide for establishing drainage slopes. The procedure for paving with a sand base is as follows:

1. Stake out the site and excavate deep enough to allow a 1″ to 2″ sand bed beneath the brick. The ground should be firm; brick laid on recently filled dirt will have a tendency to sink in spots. For

Sweeping sand into the joints

good drainage, the bricks should be kept 1″ above grade.
2. Edge the site using brick or wood framing.
3. Pour sand to a depth of 1″ to 2″, then wet the sand with a fine spray to settle it. Level it by screeding.
4. In particularly damp areas, it's a good idea to underlay the sand base with a 1″ to 2″ layer of gravel or crushed stone to improve drainage.
5. Begin laying the bricks outward from a corner, using a mason's line or string to align the rows. Tap and level the bricks in place with a mallet.
6. Sweep sand into the joints.
7. Dampen the top of the paving with a fine hose spray to settle the sand.
8. Sweep in additional sand as needed to fill the joints.

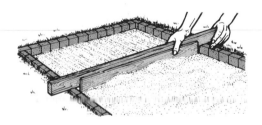

Leveling the sand base with a screed

Make sure the bricks are perfectly level.

PAVING USING THE DRY MORTAR SYSTEM

Dry mortared paving begins in a manner identical to its mortarless counterpart. After staking out the area, excavating, and placing the edging, proceed as follows:

1. Again start in a corner with the bricks. They must be set in place with 1/2″ joints; use a 1/2″ wooden spacer and a mason's line for alignment.

Spacers help in the placement of the bricks.

2. Mix dry cement and sand in a 1:3 ratio. Spread it over the surface, brushing it into the open joints. You can use the spacer to tamp the mix for firmer bonding.
3. Kneel on a board to avoid disturbing the paving. Carefully sweep any excess mix off the paving surface.
4. Keep in mind that any mix left on the surface can leave stains. However, some degree of staining is usually unavoidable with the dry mortar system.
5. Using an extremely fine spray so as not to splash the mix out of the joints, wet down the paving. Don't allow pools to

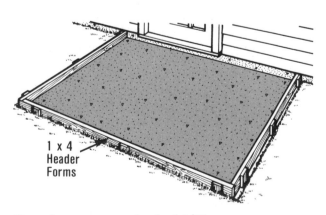

Hose down the bricks with a very fine spray.

form, and don't wash away any of the mortar.
6. Over the next two to three hours, wet the paving periodically to keep it damp. Tool the joints as desired when the mortar is firm enough. For a professional look, use a wooden dowel or broom handle for the tooling.
7. After a few hours, scrub the brick face with a burlap sack to remove stains and spillover.

PAVING USING THE WET MORTAR SYSTEM

The wet mortar system is used over a concrete slab to provide a firm, secure paving. The slab can be either an old one or a fresh one that you've just laid. You might want to use building brick rather than paving brick in order to allow for the mortar joints. This system is ideal for use in areas where winters are harsh. Remember to wet down the bricks several hours before you start the work so that they don't absorb too much water from the mortar. After staking out the area and excavating, proceed as follows:

1. Construct the header forms with 1×4 lumber against all the open sides of the slab.

Pour the mortar to a depth of 1/2".

2. Fill the forms with prepared mortar to a 1/2″ depth.

3. You may want to use a bladed screed that rides on the edgings and extends down one brick-thickness below them. The edgings can be set for this depth plus 1/2″ to allow for the mortar bed. Screed only 10 square feet at a time.

4. Lay the bricks outward from a corner, leaving 1/2″ joints between them; a wooden spacer will come in handy for this. Use a level and line for proper placement, and remember to level in two directions.

5. Gentle taps with a mallet should be used to place the brick in final position.

6. After letting the brick set in the base for about four hours, use a small trowel to carefully work the mortar into the joints. Minimize spillage to avoid staining the bricks.

7. Let harden for a half hour, then finish the joints with a jointer, broom handle, or wooden dowel.

Working the mortar into the joints

A concave jointer can be used to finish the joints.

8. Scrub the paving several hours later with a burlap sack to remove stains and spillage.

PAVING USING PRECAST CONCRETE PAVERS

Precast concrete pavers are a practical alternative to laying concrete. You can choose from squares, rectangles, circles, and hexagons, as well as all types of interlocking patterns. Many surface types are also available, such as exposed aggregate, wrinkled, textured, and smooth. Pavers can also be made to look like brick, tile, or cobblestone. You can also cast your own custom-designed paving as outlined in chapter 8. The paving thickness you use will depend on the type of traffic the pavers will have to withstand; it's best to use thin blocks for mainly pedestrian traffic and larger, thicker ones for vehicular traffic.

In this example, the pavers are used to make a patio. Note that the installation method is no different from brick paving with a sand base. The only exception is that wood edging is used here instead of brick. After excavating the site, use the following procedure:

Interlocking and Specialty Pavers

Precast concrete pavers come in many shapes and sizes.

Nailing the edging in place.

Adding a 1/2" piece of wood to the board will enable you to screed sand.

Smooth out the gravel with a screed board.

Carefully screed the sand.

1. Install 2×6 pressure-treated edging to keep the pavers from spreading out.
2. Pour a 3-1/2" layer of crushed stone or gravel and use a notched screed board to smooth out the surface. The notches on the ends of the board should rest against the inside of the edging. Make the notch cuts equal to the thickness of the pavers plus 1/2".
3. Nail a 1/2" piece of wood onto the notch. This will enable you to screed the sand that will be poured in the next step with the same screed board.
4. To improve drainage, pour a 1/2" layer of sand and carefully screed it so that the surface is as level as possible for the blocks.
5. In order to use 8"×16"×1-1/2" pavers laid in a running bond pattern, a half-block must be cut for every other row. To cut thicker blocks, use a masonry cut-off wheel on a portable circular saw or a radial arm saw. Make a 3/8" cut in one pass down the middle of one side of the block, then turn it over and cut down the middle of the other side. Make sure the cuts are precisely aligned for a clean break. Break scored blocks in half by banging them against a hard corner along the score.

It will be necessary to break some of the pavers in half.

Fill all cracks by sweeping sand over the surface of the pavers.

WARNING

When using a radial arm saw, never pull the blade toward you when cutting; the cut-off wheel can jam and create a safety hazard. Instead, pull the blade out to the end of the arm, then cut the block by pushing the blade toward the column of the saw.

---

6. Use spacers to lay the pavers in the desired pattern.
7. If the sand layer has been leveled correctly, you shouldn't need to tamp the

Use spacers to position the pavers.

pavers down to level them. However, as a final check, you might want to use a level to check the surface.
8. Pour fine sand over the surface, then sweep it until it is clean and all the cracks are completely filled.
9. For a nice effect, you can fill large joints with contrasting gravel or grow grass in between the blocks. Fill the joints with soil, then plant seeds and fertilize as you would a lawn.

As an alternative, you can fill the paver joints with sand mix. Mix one part portland cement to two parts fine mortar sand, or use pre-packaged sand mix. Spread the dry mix over the patio, then sweep it into the joints until all are full. Remove any excess mix, then spray the entire patio surface lightly to dampen the joints. Be careful not to soak them with too much water. The filler will harden and shrink slightly below the tops of the pavers.

Maintenance for concrete paver patios is relatively simple. Unlike poured concrete, pavers don't crack in freeze–thaw cycles. If the ground has heaved up from freezing and dislodged one or more pavers, simply lift out the displaced block(s), scoop out the base, and re-lay to the original grade level. If your patio is well drained, excessive frost heaving shouldn't be much of a problem. The patio can be cleaned by hosing it.

PAVING OVER CONCRETE STEPS

You can give plain concrete steps a quick and easy facelift by adding brick paving to them. Lay out the bricks in a dry run first to check their position and fit, allowing 3/8″ between bricks for the mortar joints. This project is very straight-forward; start at the bottom riser by covering it with a row of bricks, move up to the tread and cover it, and proceed up the steps in this manner.

Unless your concrete steps already have an adequate slope, lay the bricks in mortar so that they tilt forward a little to allow for good drainage. Since steps often provide the first impres-

Tilt the bricks a bit to improve drainage.

sion of a home, you'll want to take particular care with workmanship. Tool all mortar joints inward to protect against water penetration.

Chapter 15

WORKING WITH CONCRETE BLOCK

Concrete block combines the strength and durability of concrete with the ease of construction of masonry. While poured concrete remains the most popular material for floors and slabs, concrete block is often used in foundation walls in homes and above-ground walls for garages, sheds, porches, and other structures. Concrete block is less expensive than other types of masonry, it's easier to build with than brick or stone, and it comes in hundreds of sizes, shapes, textures, and colors. Decorative concrete block has become increasingly popular for building fences, privacy screens, wall panels, and veneer. For additional information on all of the different types of concrete block available, estimating block needs, and modular planning tips, please read chapter 11.

Chapter 11 also contains vital information on mixing mortar for blockwork and brickwork. Good mortar is an absolute necessity for doing good concrete blockwork. Mortar of proper workability is soft, but with good body. It spreads readily and extrudes without smearing or dropping away.

MORTARING JOINTS

Two types of mortar bedding are used with concrete masonry: full mortar bedding and face–shell mortar bedding. In full mortar bedding, the unit webs as well as face shells are bedded in mortar. Full bedding is used for laying the first or starting course of block on a footing

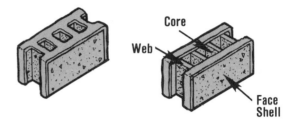

Types of mortar bedding used for laying up hollow concrete block

or foundation wall as well as for laying solid units such as concrete brick and solid block. It is also commonly used when building concrete masonry columns, piers, and pilasters that will carry heavy loads. Where some vertical cores are to be solidly grouted, such as in reinforced masonry, the webs around each grouted core are fully mortared. For all other concrete masonry work with hollow units, it is common practice to use only face–shell bedding. Also, the head (vertical) joints of block having plain ends are mortared only opposite the face shells.

For bed (horizontal) joints, all concrete block should be laid with the thicker part of the face shell up. This provides a larger mortar-bedding area and makes the block easier to lift.

For head (vertical) joints, you need to apply mortar only on the face–shell ends of the block. You can do this in a number of ways: by buttering (mortaring) the vertical ends of the previously placed block, by buttering the vertical ends of the block to be placed next, or by buttering the block already laid and the block to be laid. The

Applying mortar to the face–shell ends of the block

last method takes a bit more time, but ensures well-filled head joints.

BASIC WALL CONSTRUCTION

All concrete block walls require a properly sized and constructed footing. See chapter 7 for details. Depending on the size and scope of the job, batter boards may be useful in locating wall position and corner locations. See chapter 4 for more information on site preparation and layout.

The following section outlines the basic procedures used in laying up hollow concrete block. Solid concrete brick or block is laid up using brickwork techniques outlined in chapter 12. Decorative screen block often requires special techniques that are covered later in this chapter.

Hollow blockwork follows the same general principles of good brickwork. As always, the first course is extremely critical. It must be precisely positioned on the footer with all corners absolutely square. Alternating course designs are used to create staggered vertical (head) joints unless a continuous control joint is needed. Alternating course layouts are created through the use of half-length units or alternating corner layouts.

Once the first course is constructed, wall ends or corners are built to a four to six course height and the area between these leads is filled in with block.

First Course Construction

Follow these steps for the first course:

1. Locate the corner position on the footer, and lay out the block without mortar. This dry run is extremely helpful in checking dimensions and positioning. Remember each vertical mortar joint will measure 3/8".
2. Snap a chalk line on the footing or stretch a mason's line between corners to finalize the first course position.
3. Remove the block and spread a full mortar bed for several feet at the first corner position. Furrow the bed with a trowel as shown. It's important to have plenty of mortar along the bottom edges of the block for the first course.
4. Carefully lay the first corner block in position. The thick end of the face–shell should face up to provide a larger mortar bedding area for the second and all subsequent courses.

Lay out the first course without mortar to check block positioning and corner locations.

Preparing the full mortar bed for the first course

Positioning the corner block and first course blocks

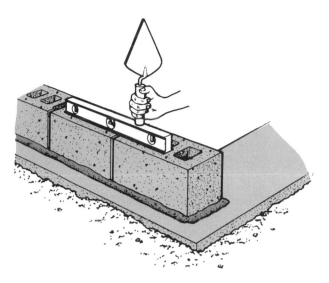

Check the first few blocks for level and alignment.

has stiffened will break the mortar bond, weakening the wall and allowing rain to penetrate.

6. After three or four blocks have been laid, use your mason's level as a straight-edge to make sure the blocks are in correct alignment.

7. Bring the blocks to proper grade and plumbness by tapping them with the trowel handle while observing the level reading. Mortar joint thickness should be 3/8". It is very important that the first course is laid with care and is properly aligned, leveled, and plumbed. This will make placing succeeding courses easier and result in a true vertical wall.

Corner Construction

As mentioned earlier, once the first course is in place, the second and subsequent courses normally use face–shell mortaring. However, in some localities, a full mortar bed is used for all courses. Be sure to consult your local building codes.

The ends of the wall are usually built first to a height of four to six courses. For simple, free-standing walls with no 90° corners, the second course is started using a special half-length unit.

5. Butter the vertical ends of the next two or three blocks and place them in position on the mortar bed. As each block is brought into final position, push it down into the mortar bed and against the previously laid block. Be as accurate as possible in your first attempt to position each block. This eliminates excessive shifting and movement of the block during placement. Don't attempt to move or straighten a block in any manner once the mortar has begun to stiffen. Final positioning must be done while the mortar is soft and plastic. Any attempt to move or shift the block after the mortar

Face–shell mortaring in preparation for laying the second course.

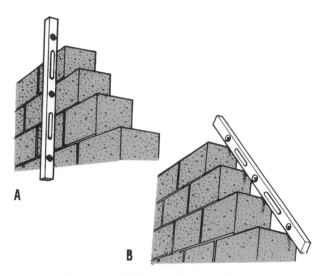

(A) Checking that all block is laid in the same plane, and (B) checking the horizontal spacing of half-step construction

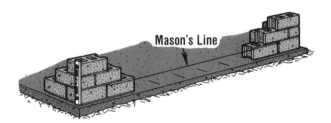

Typical 90° corner layout

This creates the necessary staggered vertical joint spacing. The third course pattern is identical to the first course, the fourth course identical to the second course, and so on.

When a 90° corner is needed at the ends of the wall, the exact layout of the blocks for alternating courses depends on block size. The reason for this is to maintain a 4″ or 8″ modular spacing to minimize the number of cut block needed. Eight-inch-thick block poses no problem. Corners are made using simple butt joints in an alternating pattern to create a staggered vertical joint spacing. Corner layouts for 4″, 6″, 10″, and 12″-thick walls may require cutting units, special "L"-shaped corner blocks, or special fill-in blocks. Modular corner layouts are illustrated on page 174.

If needed, cut blocks to size by scoring them on both sides with a brick set and then striking the waste side of the cut to break off the excess. For fast, neat scoring of numerous units, a power saw equipped with an abrasive masonry blade is the best method.

Using a story pole to check course height

As each course is laid at the corner, use your mason's level to check it for alignment, level, and plumb. Make certain that the faces of all block are in the same plane and check the horizontal spacing of each block by placing the level diagonally across the corners of the block as shown.

Finally, use a story pole to accurately check the position of the top of each course. Maintain a 3/8″ mortar joint between all courses.

Filling in the Leads

When filling in the wall between the end or corner leads, stretch a mason's line from corner to corner for each course. The top outside edge of each block is then laid to this line. Different devices can be used to fasten the line, such as a corner block held in place at the corner by tension on the line. Other options include a line pin driven into a mortar joint that has set; a line twig held by a brick to eliminate the sag in the line; and a line stretcher fitted over the top of the wall at any convenient place.

The manner in which you handle and grip the blocks is important. Practice will determine the best way for each individual. The recommended method in many quarters is to tilt the block slightly toward your body. This makes it possible to see the upper edge of the course below so you can place the lower edge of the block directly over it. By rolling the block slightly to a vertical position and shoving it against the adjacent block, it can be positioned with minimum adjustment. Remember that all adjustments must be made while the mortar is still soft and plastic.

By tapping lightly with the trowel handle, you can level and align each block to the mason's line. The use of the level between corners is lim-

Excess mortar extruded at joints should be cut off with a mason's trowel before it begins to set up.

ited to checking the face of each block to keep it lined up with the face of the wall. To assure a good bond, the mortar should not be spread too far ahead of the actual laying of block or it will lose its plasticity. When steel joint reinforcement is required, lay it on top of the block and trowel mortar over it.

As each new block is laid, excess mortar extruded from the joints must be cut off with the trowel. It may be thrown back on the mortarboard and reworked into the fresh mortar. If you're working rapidly, the extruded mortar cut from the joints can be applied to the face shells of the block just laid. Should there be any delay long enough for the mortar to stiffen on the block, the mortar must be removed and reworked. Applying mortar to the face shell of the block already in the wall and to the block being set ensures well-filled joints. Mortar that has been dropped onto the floor or ground should be discarded.

Closure Blocks. When installing the last (or closure) block in each course, all edges of the opening and all four vertical edges of the closure block should be buttered with mortar and the closure block carefully lowered into place. The mortar joint is then dressed with the point of the trowel. If any of the mortar falls out, leaving an opening in the joint, the closure block must be removed, fresh mortar applied, and the operation repeated.

Laying block to the line stretched between the corner leads

Laying the closure block. All four edges of the block are buttered with mortar.

Tooling Joints. Weathertight joints and neat appearance of a concrete block wall are dependent on proper tooling; that is, compressing and shaping the face of the mortar joint. The mortar joints should be tooled when the mortar in the section of wall just laid is hard enough so a thumbprint barely shows. Tooling compacts the mortar and forces it tightly against the masonry unit on each side of the joint. Proper tooling will produce uniform joints with sharp, clean lines. Unless otherwise specified, all joints should be tooled either in a concave shape or a "V" shape.

The jointer used for tooling horizontal joints should be longer than a block and turned up at one end to prevent gouging the mortar. Tooling of the horizontal joints should be done first, followed by striking of the vertical joints with a

Tooling mortar joints to produce a weathertight seal

small "S"-shaped jointer. After the joints have been tooled, any excess mortar should be trimmed off flush with the face of the wall with the trowel and then dressed with a burlap bag or brush.

CONTROL JOINTS

Control joints are needed in large masonry walls so that movements caused by various types of stress can occur without damaging the concrete block. A control joint is a continuous vertical mortar joint designed into the wall by combining half-length and full-length block at that point. Stresses are dissipated at the control joint, eliminating cracking in surrounding block.

There are a number of types of control joints built into concrete masonry walls. The so-called Michigan type of control joint uses conventional flanged units. A strip of building paper is curled into the end core covering the end of the block on one side of the joint and, as the block on the other side of the joint is laid, the core is filled with mortar. The filling bonds to one block, but the paper prevents bond to the block on the other side of the control joint. Thus, the control joint permits longitudinal movement of the wall while the mortar plug transmits transverse loads.

The tongue-and-groove type of control joint uses special units that are manufactured in sets consisting of full- and half-length units. The tongue of one special unit fits into the groove of another special unit or into the open end of a regular flanged stretcher. The units are laid in mortar exactly the same as any other masonry units, including mortar in the head (vertical) joint. Part of the mortar is allowed to remain in the vertical joint to form a backing against which weather-resistant caulking can be packed. The tongue-and-groove units provide excellent lateral stability for the wall.

Preformed rubber gasket strips set in the grooves of steel sash blocks can also be used to form a control joint.

In still another type of control joint, regular open-end units are tied with a Z-bar across the

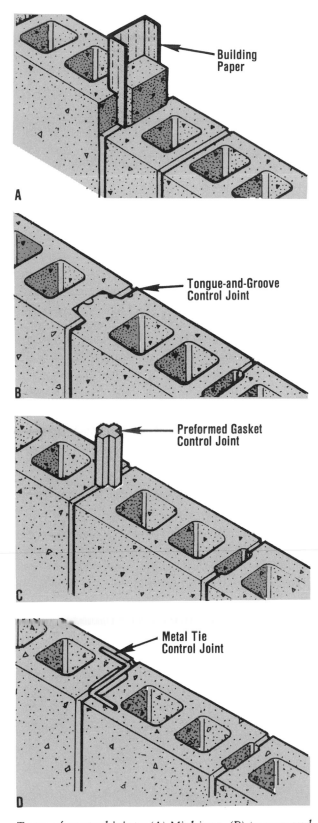

joint. All of these control joints are first laid up in mortar just the same as any other vertical mortar joint. However, if a control joint is to be exposed to view or the weather, the mortar should be permitted to become quite stiff before a recess is raked out of it to a depth of about 3/4". The mortar remaining in the control joint forms a backing to confine a caulking compound or similar elastic weathertight material. First, however, to prevent absorption of oils from certain caulking compounds, the side faces of the raked joint should be primed with shellac, aluminum paint, or other sealer, but the inner face of the joint should be greased or given some other bond-breaker. Then the caulking is applied by using a caulking gun or, for knife-grade compound, a pointing trowel. Care must be taken not to smear the caulking onto the face of the wall.

Spacing and Location

Control joints are normally located every 25' to 30' in larger structural walls. Local building codes should be consulted for control joint specifications for all concrete blockwork involving larger structures such as foundation walls, garages, and outbuildings.

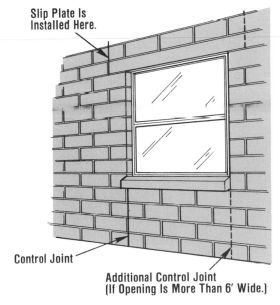

Types of control joints: (A) Michigan, (B) tongue and groove, (C) gasket, and (D) Z-bar

Windows are a natural location for control joints in concrete block walls.

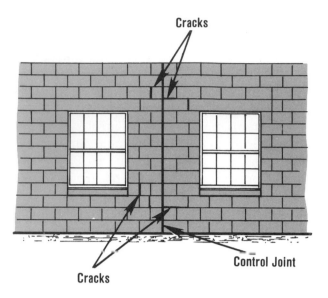

Improper position for a control joint

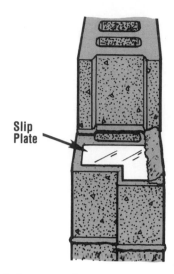

A thin metal plate provides a slip surface for the window lintel.

Obviously, control joint locations must be selected during the planning stages of the project. All large openings in walls should be recognized as natural and desirable control joint locations. Although some adjustment in the established joint pattern may be required, it is effective to use vertical sides of wall openings as part of the control joint layout. Under windows the joints usually are in line with the sides of the openings. Above doors and windows the joints must be offset to the end of the lintels. To permit movement, the bearing of at least one end of the lintel should be built to slide.

Openings less than 6' wide require a control joint along one side only, but openings of more than 6' should have joints along both sides. A control joint between two windows should be avoided since it will not function properly.

LINTELS AND SILLS

Precast concrete lintels are available for door and window openings in a variety of shapes and types to suit the wall or load they must carry. A basic type is the solid precast lintel; another is the "U"-shaped lintel or lintel block. You can use these for either metal or wood doors. A noncorroding metal plate placed under the ends of the

lintel where control joints are located will permit the lintel to slip and the control joints to function properly. A full bed of mortar should be placed over the plate to uniformly distribute the lintel load. After the mortar in the control joint and around the lintel has hardened sufficiently, it should be raked out to a depth of 3/4" and filled with a caulking compound.

Precast concrete sills are usually installed after the wall has been finished. Joints at the ends of the sills should be tightly filled with mortar or raked and filled with caulking compound. Concrete masonry sill units are sometimes available in a nominal length of 8". To ensure watertightness with these units, the course shown at

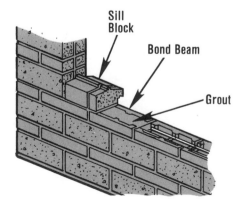

Precast sill block

the bottom of the opposite page should be a bond beam filled solid with grout or flashing provided below the sill.

BOND BEAMS

Bond beams are reinforced courses of block that bond and integrate a concrete masonry wall into a stronger unit. They increase the bending strength of the wall and are particularly needed to resist high winds and hurricane and earthquake forces. In addition, they exert restraint against wall movement and thus reduce the formation of cracks. They may even be used at vertical intervals instead of vertical reinforcing steel.

The bond beams are constructed with special-shaped block filled with concrete or grout and reinforced with embedded steel bars. Bond beams are usually located at the top of walls to stiffen them. Since they have appreciable structural strength, they can be located to serve as lintels over doors and windows. When bond beams are located just above the floor, they act to distribute the wall weight (making the wall a deep beam) and thus help avoid wall cracks if the floor sags. Bond beams may also be located below a window sill.

Reinforcement of bond beams must satisfy structural requirements but should not be less

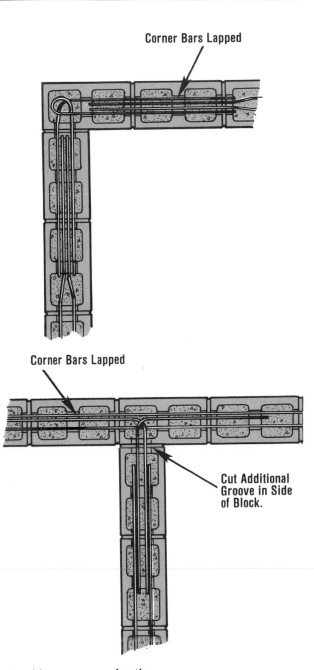

Bond beam corner details

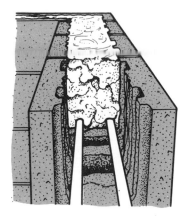

Bond beams have a solid base and allow for easy positioning of steel reinforcement. When grouted, they form a solid horizontal beam that adds greatly to the strength of the wall.

than two No. 4 steel bars in 8"-wide bond beams and two No. 5 steel bars in 10" and 12"-wide bond beams. In some earthquake-prone areas building codes require a 16"-deep bond beam with two additional No. 4 bars located in the top of the beam. Bars should be bent around corners and lapped according to the local building code.

When the bond beams serve only as a means of crack control, they should be discontinuous at

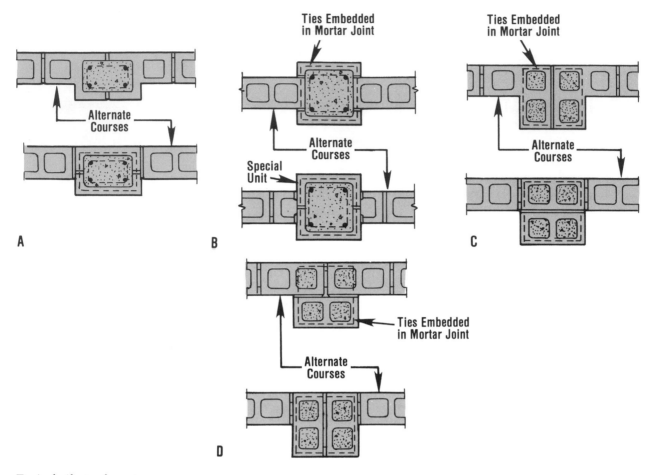

Typical pilaster layouts

control joints. Where structural considerations require that bond beams be continuous across control joints, a dummy groove should be provided to control the location of the anticipated crack.

PILASTERS AND PIERS

Pilasters are columns of thickened wall sections built as an integral part of the concrete block wall. They are used to provide additional load bearing and lateral strength, particularly in long running walls. Pilasters may project on one or both sides of the wall.

Piers are isolated columns of masonry block used to support heavy, concentrated vertical roof or floor loads.

As shown, pilasters and piers can be constructed of special block units or units similar to those used in surrounding walls. Hollow units may or may not be grouted and contain embedded reinforcement. Grouted piers and pilasters require ties embedded in the face–shell mortar bedding.

INTERSECTING WALLS

An interlocking masonry bond is not used for an intersecting bearing wall unless the intersection occurs at an outside corner. In all other cases, the end of one wall butts against the face of the second wall. A control joint is located at this intersection. Special 1/4" metal tie bars are included in the joint to tie the walls together. The

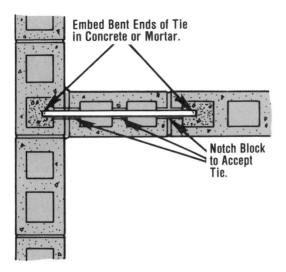

Metal bar tie details at load-bearing block wall intersections

Wire mesh ties for nonload-bearing wall intersections

tie bars have right-angle bends at each end that are embedded in block cores filled with grout or concrete. Tie bars are not used for every core. Normal vertical spacing for tie bars is 4', or about every sixth course.

For nonload-bearing walls, metal lath or hardware cloth can be placed across the joint in alternate courses.

Wood sill anchoring detail

ANCHOR BOLTS, SILLS, AND CAPS

In the construction of concrete block walls with wood framing, wood plates are fastened to the tops of the walls by anchor bolts 1/2" in diameter and 18" long, and spaced not more than 4' apart. The bolts are placed in the cores of the top two courses of block; the cores are then filled with grout. Pieces of metal lath are placed in the second horizontal mortar joint from the top of the wall and under the cores that will be filled to hold the concrete grout or mortar filling in place. The threaded end of the bolt should extend about 3" above the top of the wall. When the grout or filling has hardened, a wood plate can be securely fastened to the wall.

Four-inch-thick solid masonry block can be used to top off a concrete block wall. Special top blocks having a top web that is solid concrete, 4" thick can also be used. Local building codes may not allow the use of solid capping blocks in foundation construction, so check with local authorities for accepted treatments in your area.

COMPOSITE WALLS

In a composite wall, two wythes of different materials are bonded together with masonry joints, with each wythe contributing to the overall strength of the wall. In contrast, veneered walls consist of two separate walls tied together with anchors but having no shared masonry bonds.

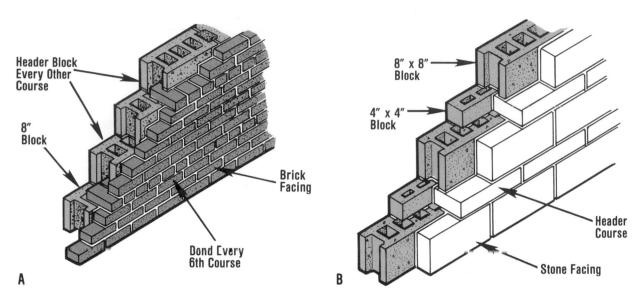

Concrete block composite wall (A) brick face and (B) cut stone

The most popular composite walls consist of a concrete block backing wythe with a face wythe of brick, concrete block, or stone. Courses of concrete block alternate between standard and special header units. When brick is used as facing, every seventh brick course is a header course.

In composite walls, the first wythe laid is parged (plastered) with mortar not less than 3/8" thick before the adjacent wythe is laid. This helps bond the two wythes together and produces an effective moisture barrier against water penetration. Before parging, however, mortar protruding from the joints of the first-laid wythe should be cut flush while the mortar is still soft; otherwise, parging over the hardened mortar may break the bond in the mortar joints and result in a leaky wall.

Facing or backing wythes can be laid first. For less experienced workers, laying the heavier backing wythe of block is better as it provides a stronger base for applying the mortar. Still, do not push too hard during parging or joint bonds may be broken.

In composite construction, all solid facing units should be laid with full mortar bedding and the head joints completely filled. In header courses, the cross joints should also be complete-

Laying a thin concrete facing block against a block wall parged with an 3/8"-thick coat of mortar

Cut any extruded mortar joints (shown) flush with the surface before applying the parging coat.

ly filled; that is, mortar is spread over the entire side of the header unit before it is shoved into place.

PATCHING AND CLEANING

Any patching of mortar joints or filling of holes should be done with fresh mortar. Particular care is needed to prevent smearing the mortar onto the surface of the block. Once hardened, embedded mortar smears can't be removed, nor can they be covered up with paint. Since concrete block walls should not be cleaned with acid, you should make it a point to keep the wall surface clean during construction. Any mortar droppings that stick to the block should be allowed to dry before removal with a trowel. The mortar may be smeared into the surface of the block if it's removed while soft. When dry and hard, most remaining mortar can be removed by rubbing with a small piece of block, then brushing the spots.

Building paper or tarpaulins should be used to cover the wall if you have to stop the work for any length of time. This will prevent moisture from entering the cores and cavities. Planks laid on the wall are not adequate cover. All bags of cement should also be covered. Keep in mind that a freshly built, freestanding wall must be braced to prevent collapse from wind or other forces.

DAMPPROOFING AND INSULATING

When concrete block is used to build foundation and structural walls for basements and buildings, two other considerations come into play: dampproofing and insulating.

Dampproofing Walls

Portland cement-based masonry coatings designed to provide waterproofing can be used on both the interior and exterior surfaces of the wall. See chapter 9 for application details.

You can also mix your own portland cement plaster (one part portland cement to two and one-half parts fine sand) or apply the same mortar mix used to lay up the block. When waterproofing with these plasters or mortars, the exterior side of the wall should be covered with a 1/2"-thick coat of plaster, preferably applied in two layers.

The wall surface should be clean and dampened by spraying with water (but not soaked) before the application of the plaster. This will prevent the block from absorbing excessive water from the plaster and will ensure a better bond. When the plaster is applied in two coats, the first coat should be troweled firmly over the masonry. When it's partially hardened, the surface should be roughed up to provide a good bond for the second coat. The first coat should be kept damp and allowed to harden for at least twenty-four hours before you apply the second coat. Extend the plaster from 6" above the finished ground line down to the footing; the plaster can be shaped to prevent water from collecting where the wall and footing meet.

Just before application of the second coat, the roughened surface should be lightly dampened with water for a good bond. Moist-cure the second coat for at least forty-eight hours after application. If a single 1/2"-thick coat of plaster is used, the surface preparation is the same. The plaster should be kept moist for at least forty-eight hours after application.

In poorly drained, wet soils, the plaster coating should be covered with two thick cement–sand grout coats scrubbed into the surface with a stiff-bristle brush. Dampen the plaster before applying the first coat of grout, then dampen the surface again before applying the second coat. Each coat of grout should be cured by being kept wet for at least twenty-four hours.

Two coats of cement-based paint, designed specifically for waterproofing foundation walls, can be used instead of the grout. In order for any dampproofing to be effective, the control joints in the walls must be sealed to keep out groundwater. Except in localities where the climate is dry or where the subgrade is well drained, a line of concrete drain tile should be placed around the outside of the footing and connected to a suitable outlet. Joints between tiles should be protected with pieces of building felt. Cover the tile with at least 12″ of coarse gravel or crushed stone prior to backfilling.

Insulating Walls

Heat loss through concrete block basement and structural walls can be dramatically reduced through the use of insulation. Various insulation methods include:

- Granular lightweight fill poured into the cores as the wall is built
- Blocks manufactured with a layer of insulation inside
- Blocks of lightweight aggregate that provides better insulating properties
- Rigid insulation board applied to the exterior wall surface and then plastered or coated to create the wall's final finish (see chapter 9, page 147, for a detailed explanation of this insulating method)

SCREEN WALLS

Walls, fences, and partitions constructed of concrete masonry screen block offer privacy while at the same time providing a degree of sunshine and airflow control.

Your main concern in planning and constructing a screen wall project is stability and strength. Most screen block designs are laid up using continuous vertical mortar joints that are weaker than staggered joints used in standard blockwork. The open, airy design of screen block also reduces its overall strength and load-bearing capability. Without additional reinforcement, screen walls can become unstable in high winds, or fail due to natural expansion and contraction.

Fortunately, in most home and garden applications, screen walls are required to bear no

more than their own weight. In fact, many codes prohibit the use of screen block in load-bearing applications. Nonload-bearing screen walls require varying amounts of vertical and horizontal support based on wall height and length, screen block thickness and design, and local weather and wind conditions. The examples on the next page are general in nature. Consult your building supplier for the preferred method of supporting the block design you select and check all building code regulations before beginning any screen block project of appreciable size.

Low decorative walls up to a height of 3′ or so usually only require welded steel joint reinforcement in the bed joints of every other course. This horizontal reinforcement should stop just short of the ends of the wall. Press the reinforcement into the mortar bed prior to placing the lead block of the course. As you fill between the leads, trowel additional mortar onto the steel reinforcement and lift it slightly as you set each block in position. This helps embed the steel at the center of the bed joints. Bed joint thickness should be a minimum of 3/8″ or double the thickness of the reinforcement used.

Screen wall higher than 3′ generally requires additional lateral support to supplement the bed joint reinforcement. Lateral support is normally provided by reinforced concrete block pilasters or vertical steel channels built into the wall at properly spaced intervals. Screen blocks tie into the fully grouted pilasters using steel reinforcement. As shown, vertical steel channels

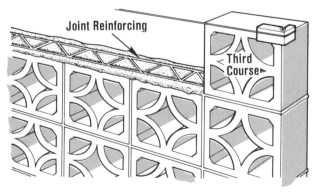

Welded steel joint reinforcement is sufficient for low screen walls.

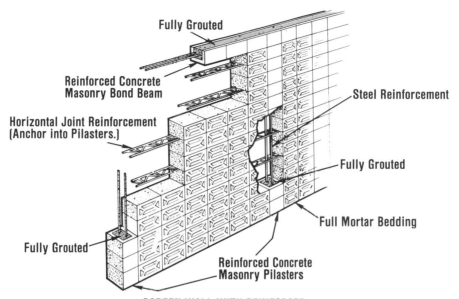

Fully Grouted

Reinforced Concrete
Masonry Bond Beam

Horizontal Joint Reinforcement
(Anchor into Pilasters.)

Steel Reinforcement

Fully Grouted

Fully Grouted

Full Mortar Bedding

Reinforced Concrete
Masonry Pilasters

**SCREEN WALL WITH REINFORCED
CONCRETE MASONRY STRUCTURAL FRAME**

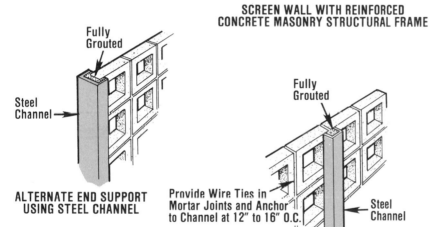

Fully
Grouted

Steel
Channel

**ALTERNATE END SUPPORT
USING STEEL CHANNEL**

Fully
Grouted

Provide Wire Ties in
Mortar Joints and Anchor
to Channel at 12" to 16" O.C.

Steel
Channel

**ALTERNATE INTERMEDIATE SUPPORT
USING STEEL CHANNEL**

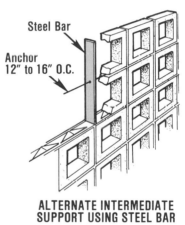

Steel Bar

Anchor
12" to 16" O.C.

**ALTERNATE INTERMEDIATE
SUPPORT USING STEEL BAR**

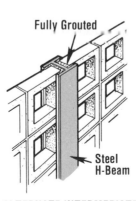

Fully Grouted

Steel
H-Beam

**ALTERNATE INTERMEDIATE
SUPPORT USING STEEL H-BEAM**

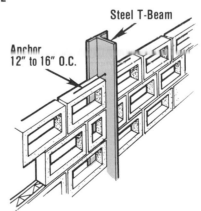

Steel T-Beam

Anchor
12" to 16" O.C.

**ALTERNATE INTERMEDIATE
SUPPORT USING STEEL T-BEAM**

Lateral support details for tall screen walls

are also fully grouted or tied into the adjacent screen block with metal ties. Steel beam channels and pilaster rebar must be embedded in (or properly anchored to) a properly sized and constructed footer.

Work carefully to ensure full, consistent mortar joints on all horizontal and vertical surfaces of each block. Joints should be tooled to be as watertight as possible. In addition, when hollow masonry units are laid with their cores vertical, the top course should be capped to prevent the entrance of water into the wall interior.

MORTARLESS BLOCK CONSTRUCTION

It is also possible to construct concrete block walls without the use of bed and vertical mortar joints. In this type of mortarless block construction a portland cement-based blend of fiberglass and fine sand is applied to both surfaces of dry-stacked block. The mixture is troweled onto the surface to a thickness of approximately 1/8". It leaves an attractive, waterproof, stucco-like finish. Only the first course uses a bed joint to bond it to the footer.

Mortarless dry-stack block construction using a 1/8" coat of fiberglass-reinforced bonding cement

It's literally a "stack and stucco" project. A single coat of this special surface-bonding cement provides greater flexural and impact strength than mortar joints. This type of construction also compares very favorably economically with standard mortared construction.

In planning the project, remember that the actual rather than the nominal size of the block must be used since there will be no 3/8" joints. The bed joint for the first course on the footer is 1/8" thick. For a tight-fitting wall and good bonding, make certain the blocks are free of dirt, soil, and grease, and remove any burrs and chips on the block with a chisel and hammer.

The wall-bonding cement is a specialty product available from several manufacturers. All products work in basically the same manner and are applied as outlined below. However, check the premix manufacturer's specific instructions for mixing, application tips, and construction details.

1. To ensure square, accurate corners and a straight wall, dry-lay the corner leads and first course and chalk-mark the block positions on the footer. Use a level and mason's line for accuracy.
2. Remove the blocks and lay a 1/8" bed of mortar on the footer.
3. Re-lay the bottom course, checking alignment and level every three to four blocks. It's important that the bottom course is laid properly, since all the succeeding courses will be stacked on top of it.
4. For projects with wall openings or intersecting walls (other than at corners), refer to steps 6 and 7. Otherwise, dry-stack the remaining block in a running bond pattern to the desired height. If necessary, shim with sand or dry mortar mix. Stack block to create continuous vertical control joints at required locations (see page 226).
5. Wet down the wall and apply a 1/8" coat of the cement–fiberglass mix to the sides and exposed ends of the wall. Use a finishing trowel with an upward motion.

Apply the coating with a trowel in an upward sweeping motion.

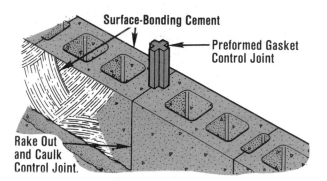

Control joint design for mortarless construction is the same as that for standard mortared work. Always rake the wall coating from the control joint.

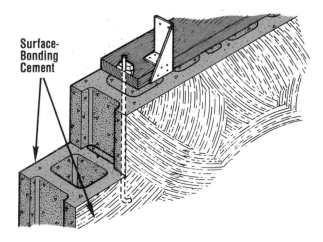

Capping sill detail for mortarless block construction that can also be used in standard construction

6. Stack and bond the main and bearing wall intersection together in four- to six-course intervals. After the main wall has been stacked and coated no higher than six courses, begin the intersecting wall. It also should be placed in an 1/8" bed of mortar. Embed metal ties every four to six courses. Stuffing the bottom of the block opening with paper will allow you to fill it with mortar to hold the tie. Continue to build the intersection in intervals to the full height of the wall.

7. For window or door openings, dry-stack the block no more than two or three courses higher than the bottom of the opening before framing out the opening to the exact dimensions. The wall is then built up around the frame. Precast lintels make finishing the top of doors and windows easy. Simply lay them in place and coat with the cement–fiberglass mix. Control joints are constructed as in standard block construction.

8. Regardless of the control joint design used, rake out the wall joint before the coating sets up and caulk it weathertight.

9. Attach capping sills to the top course by anchoring bolts firmly into the concrete or cement–fiberglass mix. Moist-cure the wall after eight hours by dampening with a fine spray. Repeat several times daily for three days. Roof or floor construction can proceed when the curing is complete.

MORTARLESS BLOCK PROJECTS

Once you've become comfortable working with mortarless block construction, you can use it for a wide variety of building projects, particu-

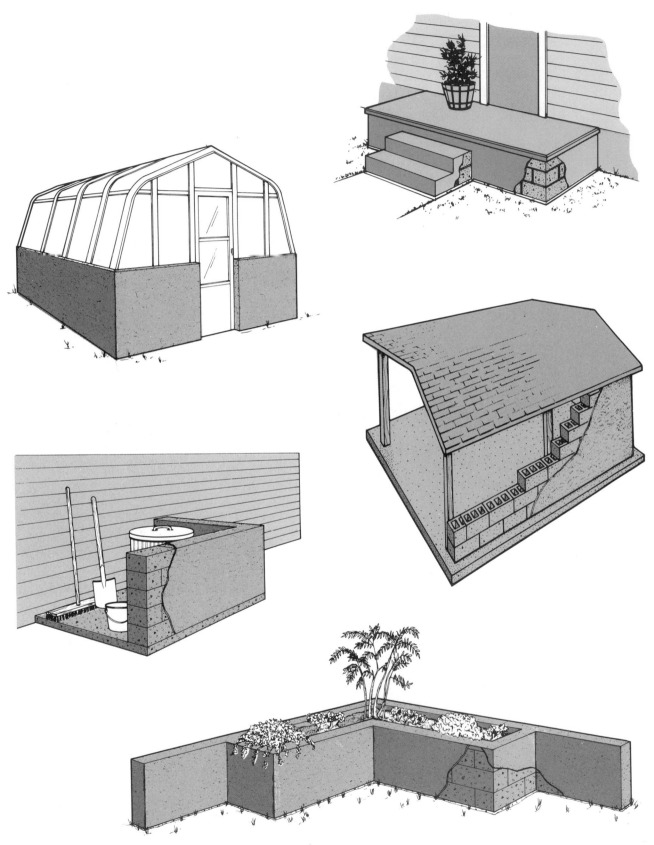

Mortarless block construction project ideas. These projects can easily be adapted to standard block construction.

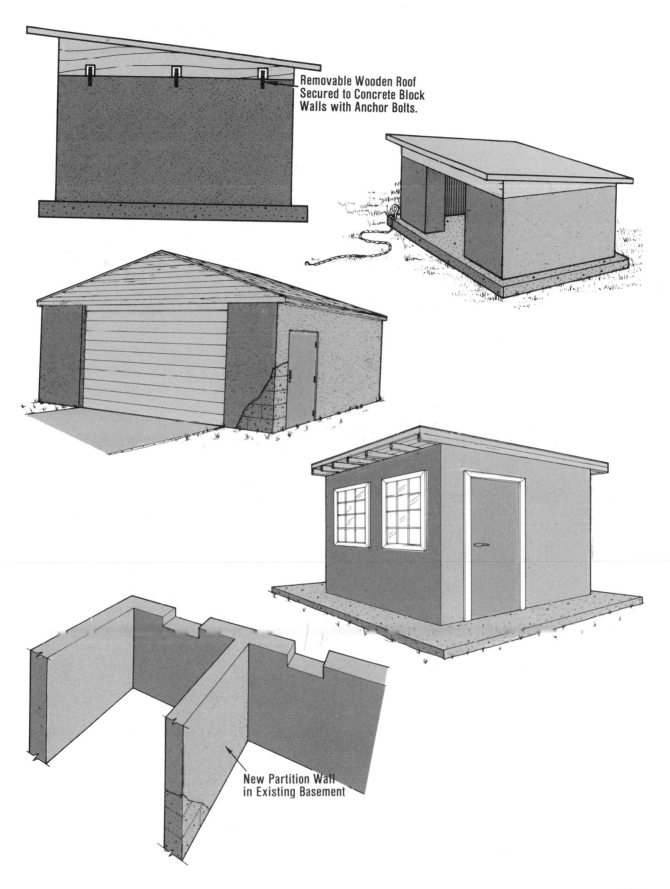

Removable Wooden Roof
Secured to Concrete Block
Walls with Anchor Bolts.

New Partition Wall
in Existing Basement

larly when a good-looking surface finish is desired. Sheds, greenhouses, carports, planters, dog houses, steps, and screen walls are several applications. Of course, the same projects could be successfully completed using standard mortar joint construction. Mortarless block is simply a very viable option to standard construction.

Mortarless Block Barbecue

A mortarless block barbecue is a relatively simple, highly useful first-time project for novices at this type of construction.

1. Excavate an area 24" wider and 8" deeper than the length and combined widths of the grill and counterboard you'll be using. Dig to a depth of about 4", keeping the edges as straight as possible.
2. Construct the footer and moist-cure it three to four days before beginning the construction.
3. Dry-stack the first course of blocks on the footer to determine their correct placement. Begin at one end, interlocking the side walls with the back wall. Mark the location of the blocks on the footer.
4. Lay a 1/2" bed of mortar on the footer and set the first course in it, beginning at the same end and butting the blocks together. Make certain that this course is set straight and level.

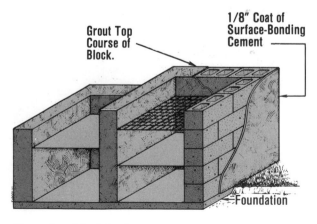

Grout Top Course of Block.

1/8" Coat of Surface-Bonding Cement

Foundation

Mortarless block barbeque

5. Stack the remaining block in a running bond pattern, interlocking all the walls together. Stack the grill walls five courses high, and the counter walls four courses high, making certain to place them plumb on the course below.
6. Place steel hangers in the grill side walls after the second, third, and fourth courses; place them in the counter side walls after the third course.
7. After thoroughly wetting down the blocks, apply a 1/8"-thick coat of cement–fiberglass mix to completely cover the wall surfaces. After eight hours, mist with a fine spray. Repeat this several times daily for two to three days to moist-cure.
8. Set the counterboard, steel plate, and grill in place on the hangers.

Mortarless Concrete Block Pool

An uncomplicated design, this mortarless concrete block pool has rectangular corners that add a dignified elegance to any garden, and its construction makes it an easy and economical project to do yourself. The combination of cement–fiberglass block coating and hydraulic water-stopping cement creates the watertight pool.

Site Excavation

1. Plan the size of your pool so that no blocks will have to be cut. Since no mortar joints will be used, use the actual dimensions of the block, which are 1/8" less than the nominal dimensions.
2. Stake out an area 1' wider on the sides and ends than the planned external dimensions of the pool.
3. Excavate the site. For a reflecting pool using only one course of blocks, excavate at least 4"; for a fish or plant pond using two block courses, excavate at least 12". For the overflow pipe to work, a minimum of 3" of the pool, including

the capping blocks, must be exposed above ground.

4. The bottom 2″ of the site are excavated only to the external dimensions of the pool; try to keep these edges straight so that the concrete slab can be poured without a form.

Pool Construction

1. Lay the reinforcing wire into the pool bed.
2. Pour the prepared concrete around the wire to a depth of 2″. Use a hook to keep the wire near the center of the slab for maximum strength.
3. Screed and float the concrete after it loses its sheen.

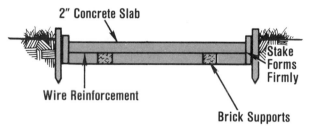

Slab details for mortarless block pool

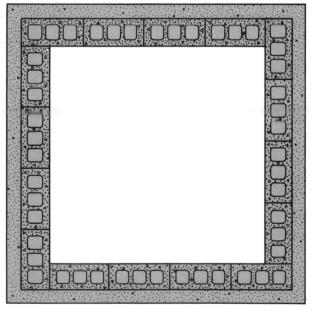

Lay block in the fresh concrete slab and "caulk" all base joints with hydraulic water-stopping cement.

4. Set the blocks lengthwise into the fresh concrete. Lay them outward from a corner, butting them together as tightly as possible and keeping them aligned with the level.
5. Appropriate blocks should be notched to locate an overflow pipe and, if used, a drainpipe. The drainpipe should exit through the wall at the base of the pool and have a flow valve or other capping device. The overflow pipe or tube is located slightly below the capping block level. In the event of a heavy rain, it will keep water from entering the block cores through the wall cap joint.
6. After the block has been set in the fresh concrete, apply a bead of hydraulic water-stopping cement along the inside and outside base of the concrete block. The hydraulic water-stopping cement creates the necessary watertight seal at the pool's base.
7. Insert any overflow and/or drainpipes and seal their openings with hydraulic water-stopping cement. Pack the openings tightly. It is important to create a watertight seal at these locations.
8. Trowel the prepared fiberglass wall coating mix onto both sides of the block wall. Take care to apply an even 1/8″ coating to all surfaces.
9. Lay a 1/4″ bed joint of fiberglass coating mix on the face–shells of the top course and lay capping blocks into position. Level the capping blocks in two directions and make sure they are correctly aligned.

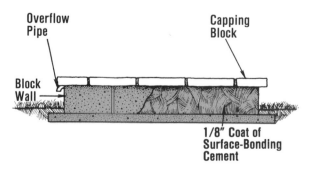

Capping block and overflow pipe details

10. Cure the concrete for at least three to four days by covering the pool with plastic sheeting.
11. After curing, apply a coat of waterproof masonry coating to all wall surfaces.

Mobile Home Underpinning

Mortarless block construction is an ideal method of finishing off mobile home installations. When building up underpinning, plan the work so that the top course of block rests flush with the bottom of the mobile home.

1. Use a plumb bob to locate the exact front of the wall and then construct a suitable footer. Build up the wall as you would any other type of mortarless block installation.
2. Mobile home underpinnings require ventilation. Special foundation vents equal in size to concrete block are available

from building suppliers. These vents are simply included in the stacking pattern during construction. They should be placed in one of the middle courses every ten to twelve blocks, depending on local building codes.
3. Slope the final grade away from the home to prevent water from collecting against the block.
4. Caulk the joint between the home and top course of block and install a drip cap to prevent water penetration.

CONCRETE BLOCK RETAINING WALLS

There are two different types of concrete block retaining walls: cantilever, or buttressed or counterfort walls. All require a properly sized and constructed concrete footer set below the frost line (see chapter 7).

Cantilever Walls

A cantilever wall has a cross-sectional shape of an inverted "T" or "L." The wall is tied into the footing with rebar that passes through the cores of the block. Rebar size and spacing depends on wall height and local code provisions, so check with local authorities before beginning the project. A typical code may specify the use of #4 rebar every 32" on center in a 4'-high wall of 8"-thick block. The same 8"-thick wall built to a 6'-height might require #6 rebar on 24" centers.

The wall may use horizontal joint reinforcement in every course or rebar-reinforced bond beams for every other course. All block cores may be filled with grout, or just those containing rebar may be filled. The first course is always laid on a full mortar bed. The remaining course can use face–shell mortaring, with the exception of webs between cores to be grouted and those not to be grouted. These specific cross webs should be mortared. As with poured concrete retaining walls discussed in chapter 7, behind wall drainage is essential. This can be provided

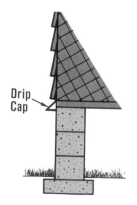

Mortarless block mobile home underpinning

Drip
Cap

Underpinning flashing detail

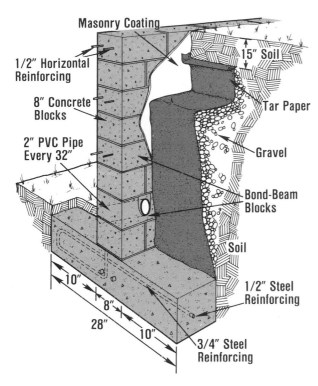

Typical concrete block cantilever retaining wall design

constructed of concrete block. When the struts are built and tied into the back of the wall and backfilled with soil, they are called *counterforts.* When they are tied into the front of the wall and left exposed, they are called *buttresses.*

CONCRETE BLOCK STOOPS AND PORCHES

Concrete block is a popular building material for constructing stoops and porches. By way of definition, a stoop is simply a smaller structure than a porch, usually less than 10'×10'. Stoops generally are not roofed over. Porches are larger versions of stoops with roof treatments that tie into the existing structure. Although similar in design, many building departments may view stoops and porches very differently with regard to property line set back and other regulations. Once your project surpasses a certain size, it will be subject to the same building code regulations that govern house construction, so check out all code guidelines in the planning stages.

For maximum stability, the stoop or porch must rest on its own footing set below the frost line. The walls of the stoop or porch should also tie into the foundation wall of the adjacent house. The method you use to meet these two design criteria depends on whether the stoop or porch is part of new or existing construction.

with through-wall drainage pipe (shown) or lateral drainage systems (see page 96).

Counterfort or Buttressed Retaining Walls

The retaining walls combine cantilever wall design with additional vertical struts or supports

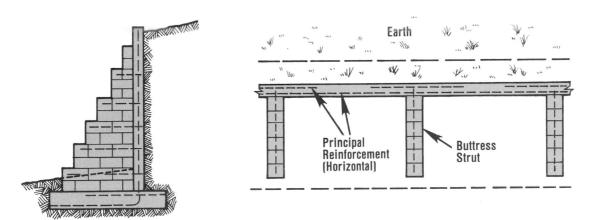

Typical buttressed concrete block retaining wall. If the buttress supports are on the back of the wall, the design is called a counterfort wall.

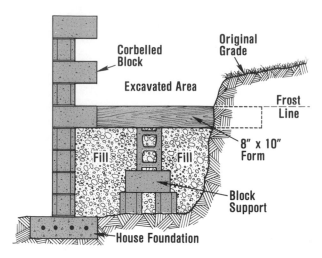

Corbelled block tie-in to house foundation for stoop or small porch construction

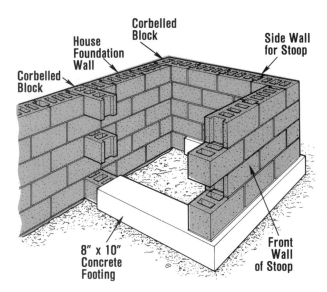

Footer and wall details for stoop or small porch construction

New Construction

Making design provisions for a stoop or porch during the construction of a home's concrete block foundation is an easy task.

The side walls of a small porch or stoop can be supported by corbelled blocks projecting at a 90° angle from the face of the house foundation wall. The first corbelled block should be located below the frost line at the level of the porch or stoop concrete footer. Alternate courses should

then be corbelled until the foundation wall reaches the height of the finished porch deck slab.

After the house foundation is complete, the porch site is backfilled to the level of the first corbelled block. Forms for the porch footer are built extending out from the first corbelled blocks, as shown. When the span between the foundation wall and firm, undisturbed soil exceeds 5' or so, support the forms with concrete block underpinning before backfilling. After the footer is placed and sufficiently cured, the walls of the stoop or porch are built using standard block construction techniques.

The corbelled block technique does not provide sufficient load-bearing support for larger porches. A large porch should have its own foundation wall that rests on the house foundation footing and is tied into the house foundation wall with steel ties.

The house footer should be widened at this location to support the porch foundation. The porch foundation wall is laid up at the same time as the house foundation wall. Footers for the porch's side and front walls must be below the frost line, but need not extend all the way down

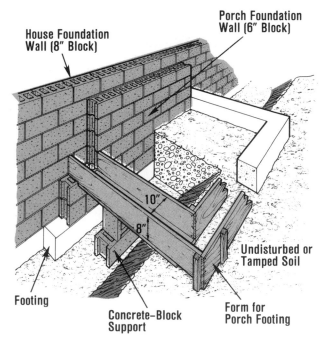

Foundation wall and footer details for larger porch construction

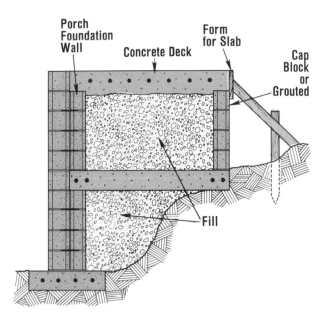

Cross-section of porch footer, wall, and concrete deck slab construction

to the house footer. As shown, a stepped shoulder design in the porch's rear foundation wall makes it easy to tie in the footer at the required frost line depth. With the footer in place, porch walls are built to the height of the finished porch deck slab.

Old Construction

Most renovation projects involve adding a new stoop or porch to an existing structure. Obviously, the corbelled construction technique for stoops and small porches cannot be used. How-

ever, you will have the advantage of a solid earth base because any backfill placed years earlier will have had time to settle.

When the ground at the site is firmly compacted, simply dig a footer trench, place the concrete footer, and construct the concrete block stoop walls.

This technique should be limited to small projects. Even the firmest soil cannot be counted on to support a larger porch without some settlement that can throw off house-to-porch alignment. For larger projects, the safest and most reliable technique is to excavate down to the existing house footer, construct a second porch footer adjacent to it, and then follow the procedure for new porch construction.

As you construct the porch's rear foundation wall on the new footer, tie every second course of block into the house's old foundation by knocking small holes into the block, inserting hooked metal ties, and mortaring the holes closed.

Deck Slabs

The stoop or porch can be capped with either a wooden deck or a poured-in-place, reinforced concrete slab. Construction of the concrete slab decking is covered in chapter 8. Use solid capping block for the top course or fill in the top course webs with grout or concrete. Use clean fill inside the walls of the stoop or porch to reduce the amount of concrete needed.

Chapter 16

MASONRY MAINTENANCE AND REPAIR

Major problems seldom occur in masonry if the brick, block, or stone is layed up correctly. Joints made of quality mortar that has been properly placed and tooled are extremely water resistant and long lasting.

Time and exposure to weather do take their toll, however, and slight mortar joint cracking and looseness will occur eventually. As in concrete repair, taking care of small problems when they appear is the best method of avoiding major failures. If left unrepaired, water will seep in between the cracked, loose mortar and cause interior wall damage and structural weakness. Make it a point to periodically inspect all masonry and concrete work on a regular basis. The best time for these inspections is early spring, when new problems associated with freeze–thaw cycles and deicers will first appear. Be especially on the lookout for water damage.

TUCKPOINTING MORTAR JOINTS

Tuckpointing is the most common repair performed on brick, block, and stone masonry. It involves cutting or chipping out old, loose mortar, and replacing it with fresh mix. Tuckpointing, or repointing as it is sometimes called, is preventive maintenance for your masonry work. It is not difficult. In fact, you may find it one of the most enjoyable repair projects you've undertaken.

Mortar for Tuckpointing

A stiffer-than-normal mortar mix is recommended for tuckpointing. You can use a just-add-water mortar premix or a mix from scratch. See chapter 11 for mix proportions when mixing from scratch using masonry cement or portland cement and lime.

A number of cement repair products can also be used for tuckpointing, such as vinyl resin patcher and hydraulic cements. If you plan to use a specialty product for tuckpointing, read the application and mixing instructions listed on the package.

When regular mortar is used for repointing, it is recommended that you prehydrate the mortar prior to tuckpointing joints. This step will greatly reduce the shrinkage of any of the joints away from the edge of the bricks and cut down on the number of hairline cracks that occur when the mortar begins to dry against the old bricks.

To prehydrate the mortar, mix the required amount of mortar mix with just enough water to form a damp unworkable mix that retains its form when pressed into a ball in your hand. Let the mortar set for about thirty to forty-five minutes, and then add enough water to make the mortar workable. The end result will be a mix slightly drier than that normally used to lay new brick. Never make large batches of mortar for repointing; mix an amount you can use in about forty-five minutes.

Cutting out loose or cracked mortar joints using a ball peen hammer and chisel

Preparing the Joints

The first step in the repointing procedure is to chip out all the loose mortar. Professional bricklayers use special abrasive wheels for this task, but you can do the job with a chisel and a ball peen hammer or a small sledgehammer. To prepare the joints, follow these steps:

1. Cut out the mortar joints to an approximate depth of 3/4″ to 1″, even if the deterioration seems to be only on the surface. This depth is needed because new mortar will not hold if it is laid in too thinly. Work carefully to avoid damaging any bricks. The use of a plugging or joint chisel will help prevent binding in the joint and chipping of the brick edges.

2. A clean surface is needed for good bonding. Rake out excess mortar or grit from the joints with a jointer tool. Then use a stiff fiber brush, vacuum cleaner, or air compressor to remove any particles of mortar or sand. Do not use a wire brush for this job. It would leave behind wire particles that would react with moisture and cause rust stains in the bricks or blocks and the new mortar.

3. Flush out any remaining particles with a garden hose equipped with a spray attachment.

Tuckpointing

To prevent the bricks from absorbing moisture out of the fresh mortar, dampen the cleaned joints with a brush and water before repointing. You could also use the fine spray of a garden hose. However, do not soak the joints so that puddles of water remain. Use the following procedure to repoint the joints:

Mortar joint that has been properly prepared for tuckpointing

Pressing fresh mortar into the joint cavity with the jointing tool

Striking the tuckpointed joint smooth with the sur-rounding joint

Chipping old mortar from the base of the removed brick

1. Load the trowel with mortar. Then pick up the mortar from the trowel with the jointer tools and press it into the joints. Pack the mortar firmly into the joints. Repoint the head joints first and the bed joints second. This sequence allows you to make unbroken horizontal strokes with the jointer to form straight, even bed joints.
2. In most cases, joints are filled flush to the wall face, then slightly depressed with the jointer and brushed clean. If concave or "V" joints were used in the original work, strike with the proper tool for these finishes after the repointing is done and before the mortar gets too hard to work.
3. To decrease the possibility of cracking or sagging in extremely deep joints, fill in about half of the joint depth, wait until the mortar is thumbprint hard, and then repoint the remainder of the joint.
4. In hot or windy conditions, periodically dampen the repointed joints to prevent the mortar from drying too fast. Spray the finished job with a fine water mist to aid in the curing process.

Repointing of stonework and block is essentially the same as brickwork. Tool the joints to match those in the sound sections of the wall.

REPAIRING LOOSE OR BROKEN BRICK

Tuckpointing replaces a portion of the original mortar joint. If the joint has deteriorated to the point where the brick is loose, or if the brick itself is damaged, the entire mortar joint should be cleaned away, and the brick should be removed from the wall and replaced or reset. Follow these steps:

1. Use the chisel to remove as much of the mortar joint as possible. You can then use a thin pry bar or end of the chisel to work the loosened brick out of its cavity. Be careful not to damage surrounding brick as you do this. Of course, if the brick is cracked or broken you can break it up completely and pull the pieces from the cavity. But if the brick is intact and reusable, do your best to remove it in one piece. Finding an exact replacement brick can be difficult.
2. If the brick is reusable, chip off all old mortar from the brick faces using a chisel or brick hammer. Remember to wear goggles and gloves during this operation. Dip the brick into a bucket of clean water to remove mortar dust and dirt and wet down all surfaces.

Chipping old mortar out of the wall cavity

The wall cavity must be absolutely free of dirt and dust.

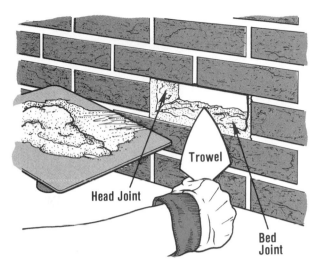

Apply a 3/4" layer of mortar to the bed and head surfaces in the cavity.

Your mason's hawk or a small section of plywood can be used to support the buttered brick as it is slid into the cavity.

3. Remove all old mortar from the cavity as well, again being careful not to damage the surrounding brick.
4. Use a stiff-bristled brush, shop or hand vacuum, or compressed air to remove all dust and dirt from the cavity. Dampen the cavity with a wet brush or the fine spray of a garden hose. This will prevent surrounding brick from drawing moisture out of the fresh mortar.
5. Use a pointing trowel to apply a 3/4" layer of fresh mortar to the bottom and sides of the cavity.
6. Butter the top of the damp replacement or refurbished brick with a 3/4" bed of mortar.
7. The easiest method of installing the buttered brick in the cavity is to place the brick on a mason's hawk or section of plywood held level with the bottom of the cavity. Simply slide the brick off the hawk or board into the cavity until the brick face is flush with the surrounding brick. Tap the brick with the handle of your trowel as needed. Mortar should extrude from the joints. If it doesn't, you did not apply enough mortar. Remove the brick, clean the area, and repeat the process using thicker joints.
8. Once the brick is in place and the joints are filled, scrape off the excess mortar

Replacing a damaged header brick

A

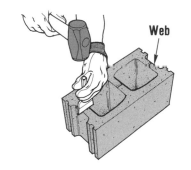

B

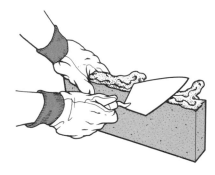

C

with the flat edge of the trowel blade and strike the joints with the appropriate finishing tools.

If the damaged brick is a header brick, it need not be entirely removed if the mortar joint is basically sound. Simply cut back the header brick at least 4″ as shown. Cut a brick to the correct replacement size, accounting for mortar joint size. Butter the bottom and sides of the cavity and the top and back of the replacement header. Install as just described.

REPAIRING CRACKED OR BROKEN CONCRETE BLOCK

Small cracks in the face of concrete block can be repaired using the techniques outlined in chapter 10 for concrete repairs. When a block is too badly damaged to patch successfully, its entire face surface should be replaced. This partial replacement is easier than removing the entire block. Cut back the mortar joints around the damaged block and break away the block face to expose the interior webs of the block. Cut the webs back about 2″ or so and clean all loose mortar and concrete block chips from the cavity.

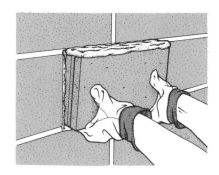

D

Steps in replacing a damaged concrete block: (A) chip back the face and webs to a 2″ depth; (B) cut a new block face from a whole stretcher block; (C) apply mortar to the face and webs; and (D) install the replacement block into the buttered cavity.

If you must purchase a new block, have the replacement face piece cut to size at the supply yard. If you plan to use a block you have on hand, you'll have to cut it to size using a brick set as shown. Score the cut line carefully, and break off small sections at a time to prevent cracking the face area.

Dry-fit the face piece into the cavity and trim the webs as needed so the face will rest flush with the surrounding block when mortared in place.

Dampen the cavity and replacement block. Apply a 1/2" layer of mortar to the bottom, sides, and webs of the cavity. Butter the top of the block as shown and fit it in place. If some of the mortar falls into the wall cavity as the block is positioned, repoint the joints after the face block is in final position. Scrape off excess mortar and then retool all joint surfaces to match surrounding treatments.

REPAIRING SHRINKAGE CRACKS

Shrinkage cracks may occur in the mortar joints of a brick or block wall. In some cases they may even split the masonry units themselves. These unsightly cracks are caused by structural shifting and settling. You should repair minor cracks immediately, before they cause major problems. Relatively minor shrinkage cracks can be filled using the tuckpointing procedure described earlier. The only difference is that the mortar joints should be cut out to a depth of at least 2" or until solid mortar is reached.

If the cracks are extremely deep or wide, the repair cannot be made by tuckpointing. The best solution is to fill the crevice with grout. You can make a grout mix by simply adding extra water to the same mortar mix recommended for tuckpointing. A grout mix flows more freely. Be conservative with the amount of extra water you add, because the grout will lack the necessary strength if it is too thin.

Use a piece of plywood or wide adhesive tape to dam up the lower portion of the crack so

Fill larger shrinkage cracks with grout.

that you can pour the grout in layers. Use a "V"-shaped trough of smooth, heavy cardboard, or a funnel to pour the grout into the crack. If necessary, use a flexible piece of wire as a tamp to fill the crack. Allow the first pour to set overnight, then repeat the procedure as often as necessary until the total height of the fissure is filled. Be sure that you tool the joints of each pour before preparing for the following one.

If the shrinkage cracks recur after the repair is made, there may be a serious structural problem, so have the wall inspected by a professional.

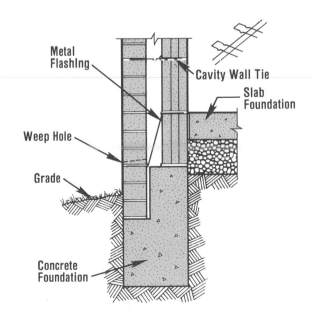

Typical cavity wall construction details using metal flashing and weep holes to direct water from the cavity hollow

CLEANING WEEP HOLES IN CAVITY WALLS

A cavity wall consists of two separate masonry walls separated by a 2" to 3" air space. The walls are bonded together with metal ties. Cavity walls increase the insulating value of the wall, prohibit the passage of water or moisture through the wall, and prevent the formation of condensation or moisture on the inner wall.

The one drawback of cavity wall design is that it is almost impossible to keep water from entering the cavity at sills, roof intersections, and so on. To channel this water out of the wall cavity, metal flashing is installed across the cavity. This flashing collects any water entering the cavity and directs it to weep holes located in the wall's outer wythe.

Weep holes are usually located in the head mortar joints and spaced every 2' or so. Weep holes are angled downward to assist the flow of water. A fine mesh plastic or metal screen is usually placed over the hole opening on the inside wall to prevent insects from entering from the outside.

Periodically check all weep holes for clogging. You should be able to insert a stiff wire through the hole and into the cavity between the walls. If the weep hole is clogged, it can be cleaned using a star drill and ball peen hammer, or a power drill equipped with a carbide-tipped bit. The star drill or bit must be slightly narrower than the weep hole diameter.

Wear eye protection during this job. Fit the tip of the star drill into the weep hole opening, angling the star drill so it will follow the angle of the hole. Remember, you do not want to enlarge the hole or change its angle. You simply want to remove any debris that has collected in the passageway.

Work the star drill into the hole by alternately striking it with the hammer and turning it one-quarter turn. Pull the star drill from the hole every third blow or so to clear out any accumulated debris. Proceed slowly and check for success with the wire. You do not want to damage the screen or metal flashing inside the wall cavity.

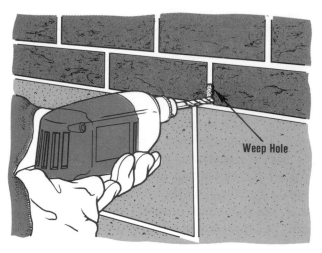

Using a carbide-tipped drill bit to clear a weep hole passage. Follow the upward angle of the hole.

When a power drill is used, the bit should be at least 4" long, and 1/16" narrower than the weep hole diameter. As with the star drill, work slowly and in stages. Reversing the drill direction once the bit has reached the cavity depth may help back debris out of the hole.

REPAIRING WALL CAPS

The top course of a brick, block, or stone wall is intended to seal the wall and prevent water penetration down into interior mortar joints. Cracks and failures along this capping

Repointing the exposed mortar joints of cap brick in garden, retaining, and other brick walls is an excellent defense against water penetration.

course will leave the wall extremely vulnerable to water damage.

Chip out cracked or deteriorating mortar joints and tuckpoint as described earlier. All loose and damaged brick should be completely removed and reset in a fresh mortar bed. Replace any damaged metal ties or flashing that you may uncover in your work.

Cast concrete caps or concrete block caps can be repaired using the patching techniques outlined in chapter 10. Badly deteriorated concrete caps can be removed and recast using the techniques outlined in chapter 8.

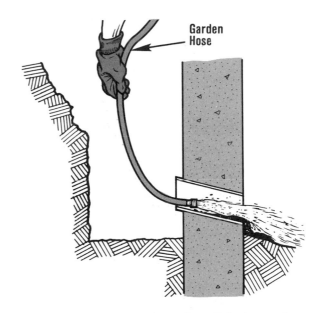

Flushing debris from a through-wall drainage pipe

CLEANING RETAINING WALL DRAINS

If your property has retaining walls, you should periodically check for proper water drainage from behind the wall. To do this, inspect the lateral or through-wall drainpipe during or just after a heavy rain. Water should exit the pipe opening. It need not be more than a steady trickle, but some sign of water drainage should be apparent.

If there is no sign of drainage, the pipe may be plugged with mud or otherwise blocked. Your first attempt should be to try and clear the pipe through the exit opening using a length of iron rod or similar tool. You should also try to flush debris from the pipe using the pressurized water jet of a garden hose.

If these efforts fail, it is likely the pipe opening behind the wall has become plugged with mud. Unfortunately, the only solution to this problem involves excavating down to the pipe location. Although not an easy job, it is considerably less trouble than repairing or replacing a buckled or collapsed retaining wall. On through-wall drainage systems proceed as follows:

1. On the backside of the retaining wall opposite the drainpipe exit, carefully remove all sod and set it aside.
2. Carefully dig down parallel to the wall until the drain opening is visible. As you near the level of the pipe, be careful not

to damage it with the edge of the shovel or pickax.
3. Clear away all soil from in and around the pipe opening.
4. Use an iron rod or similar tool to completely clear the passage. Insert a garden hose into the pipe at the rear of the wall and flush the drainpipe clear.
5. Dig a drainage area directly under the pipe opening that is about 1' square and

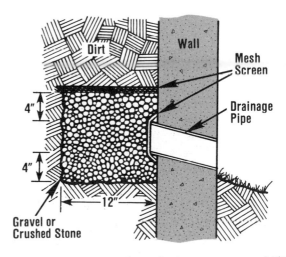

A wire mesh screen and crushed stone or gravel "filter" will prevent future retaining wall drain clogs. Use the same technique in new construction.

4″ deep. Fill the area with small crushed stone or gravel.

6. Cover the opening of the drain with a piece of 1/4″ wire mesh screening. Bend the screening so it remains in position and backfill against the screen with more crushed stone or gravel. The gravel or stone bed should extend at least 4″ above the opening of the pipe.

7. Cover the top of the gravel or stone layer with a piece of the 1/4″ mesh screening.

8. Refill the hole with soil. Be sure to tamp the soil down after each 6″ layer is added. The final grade should be slightly higher than original ground level to accommodate settling. Replace the sod and water lightly.

A clogged or faulty lateral drainage system can require a major excavation, particularly if you do not know the location of the pipe ends or how the system was laid out or designed. When this much excavation is involved, it is best to remove the old drainage line (if any) and replace it with a new line based on the system shown on page 96.

REPAIRING CHIMNEYS

Although repairing a leaning chimney or one that is leaking smoke through the brick joints are jobs best left to a professional, there are some important repairs that anyone can perform. These include repairs to preserve the integrity of the brickwork and to avoid more serious and expensive work in the future. If your chimney has crumbling pointing, or holes and cracks in the cap, doing the repairs yourself immediately will prevent further weakening and avoid the costly job of rebuilding or replacing it later. Getting up to the chimney might require some time and effort because scaffolding is usually necessary, but the repairs themselves are straightforward tasks.

Before beginning any chimney work, close all dampers to prevent any soot knocked loose by the work from entering the house. Always wear eye protection during chimney work that involves chipping away old mortar and concrete. Work gloves are a good precaution, too. Any injury that hampers your ability to see or hold on firmly can be extremely dangerous when working on scaffolding or the roof. Finally, if you are the least bit uncomfortable working at heights, have the job done by someone else. Use the following procedure for chimney tuckpointing:

1. Hold the chisel at a sharp angle to the joint while striking it with a hammer and knock out the crumbling joints to a depth of at least 1/2″. Work the chisel along the joint about 1″ at a time.

2. Chip out vertical joints first, then do horizontal joints.

3. Brush any loose particles of dirt or mortar out of the joints.

4. Dampen the joints.

5. Prepare as much mortar as can be used in one hour and trowel it into the vertical joints first, then the horizontal joints. Strike the joints flat.

6. After the mortar hardens, brush off any excess.

7. In the course of repointing, you might come across a loose brick or two that should be replaced. Use the chisel to remove the mortar from around the loose brick, then lift it out. Be careful not to damage the chimney flue. Scrape all loose or crumbling mortar from the hole.

8. After cleaning the brick of any mortar and dirt, dampen the brick as well as those surrounding the cavity. Butter its top, bottom, and both ends and press it back into the opening. Make sure the joints are completely filled and strike them flat.

Cap Repairs

Chimney caps can range from the simple sloped concrete cap shown here to more elabo-

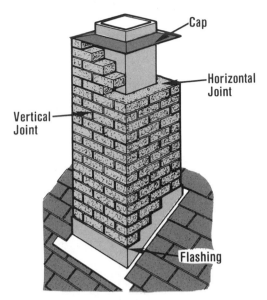

Typical brick chimney flashing, flue, and cap details. A simple sloped mortar cap is shown.

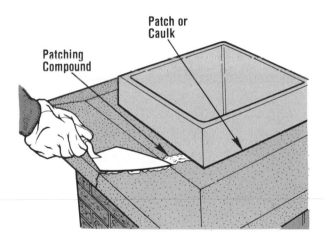

Patching a cast concrete chimney cap

rate overhanging designs that require separate formwork. Once again, patching versus replacement depends on the extent of the damage or deterioration. Use this procedure.

1. Brush out any loose particles, dirt, and organic matter from the cracks and holes in the concrete cap.
2. Dampen the crack or hole and trowel in fresh mortar or concrete patching mix. Fill the crack flush to the surrounding surface.

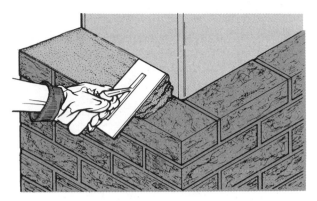

A slightly stiff mortar mix or 3:1 sand and cement mix can be used to build up a sloping chimney cap.

3. Any cracks between the cap and the flue liner should be filled with patching material. Caulking or roofing cement designed for this application can also be used.

Cap Replacement

The cap should be replaced if it is badly cracked or if the mortar is so loose it is pulling away from the brickwork and flue. Use a hammer and chisel to chip the cap away in pieces, again taking care not to dampen the flue lining. For safety, make absolutely certain no one is below when you drop the pieces to the ground.

Brush away all dirt and loose particles. For a simple sloped cap, trowel on the fresh mortar in several thick layers. Use a slightly stiffer than normal mix. Slope the cap down from the flue liner to the outside edges of the chimney stack. Press the mortar tightly against the flue to create a water-tight joint.

More elaborate cast concrete chimney caps can be constructed using simple formwork. Follow the procedures outlined in chapter 8 for new cap casting.

Chimney Flashing

Metal flashing is used to seal the roof–chimney joints. A typical flashing setup consists of two separate sections: the base and the cap.

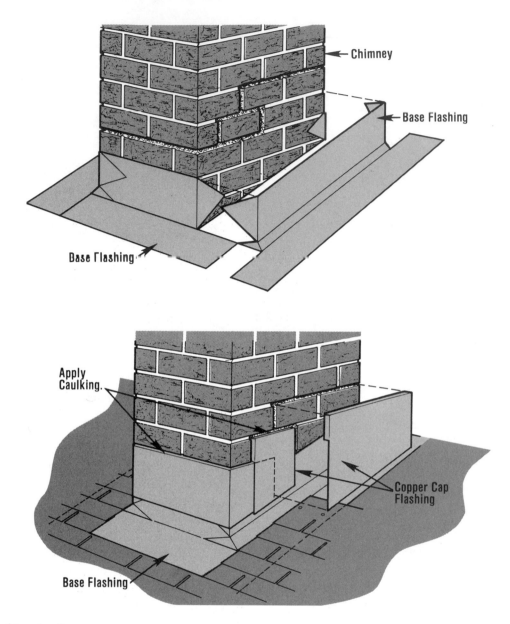

Chimney flashing details

The base section pieces are bent to conform to the slope of the roof and are positioned flush against the chimney. They extend a short distance up the chimney and a short distance out onto the roof where they are set in a bed of roofing cement, nailed in position, and covered with shingles.

The second section is the cap or counter-flashing. Cap flashing is placed over the base flashing. It extends only a very short distance out onto the roof. The top edge of the cap flashing is bent at a right angle and embedded into the mortar joints of the chimney brick.

If these joints loosen, the cap flashing may pull away from the side of the chimney, allowing water penetration. Interior water damage around the chimney perimeter is a sign this many have occurred.

If the cap flashing is still in good shape, a simple repointing is all that is needed. If the cap

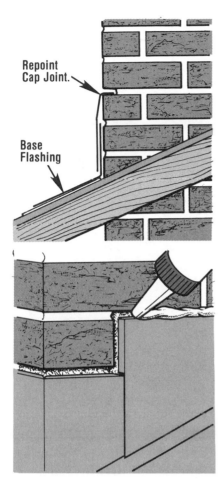

Tuckpointing the cap flashing into the chimney mortar joints

flashing shows signs of deterioration, replace it with new sheet metal. To repoint the joints:

1. Rake out all loose and crumbling mortar.
2. Push the flanged lip of the cap flashing back into the cleaned joint.
3. Pack fresh mortar into the joint on top of the cap flashing flange.
4. When the mortar has cured, caulk the cap joint.

If the base flashing is damaged or deteriorated, the entire flashing system should be replaced. This will involve the removal of surrounding shingles. Remove the shingles and base flashing carefully, noting how all components fit together and overlap. Reinstall flashing and shingles in reverse order or use the system illustrated above. Embed the base flashing and shingles in roofing cement to form water-tight joints.

REPAIRING SUNKEN BRICK OR STONE PAVING

Repairing stone or brick paving that has sunk is a task even the novice do-it-yourselfer can perform to restore both safety and attractiveness. But before beginning, try to determine the cause of the damage and remove it to avoid repeating your efforts. If water erosion has undermined the base (such as from a gutter downspout), first try removing the source or redirecting the flow. Installing a bed of gravel or a concrete base beneath the pavement and setting the pavers in a wet mortar mixture will also increase the permanence of the repairs.

Mortarless Paving

If the cause of the sinking is only ground settlement over the years, these repairs can most likely be done with only nominal expenditures of time, effort, and money. If the cause is more serious, a different paving method might be

Removing loose pavers

Setting repositioned pavers level with the existing paving

Chiseling out a broken or loose mortared paver

Sweeping sand into the paver joints

called for. To repair problems due to ground settlement, use the following procedure:

1. Remove any loose, broken, or sunken pavers. If your examination shows that adjacent pavers have not been displaced, but that settlement has occurred beneath them as well—so that they are not sitting squarely on the base—they should also be removed.

2. Fill the depression with sand up to the level of the surrounding base.
3. Set the pavers back into place. Check their level with the surrounding pavement with the carpenter's level or 2×4. If the pavers are not level, tap them gently to seat them. Replace any broken pavers.
4. Spread sand over the repaired area and sweep it into the joints. Dampen the sand with a fine spray; repeat until the joints are completely filled. For greater strength, a dry grout of one part masonry cement and four parts fine sand can be substituted for the joint sand.

Mortared Paving

Mortared joints and a concrete base make sinking infrequent in mortared paving, but when it does occur it usually indicates a more serious problem than just ground settlement. Before making the repairs after the sunken pavers are removed, check and correct any problems with area drainage. Repair the paving following these steps:

1. With a hammer and chisel remove the pavers in the sunken area. Do this as

Chipping mortar from the faces of the paver

Tooling the mortar joint to match the surrounding surface

Repositioning the paver in the surface cavity buttered on all surfaces with fresh mortar

carefully as possible to avoid breaking the pavers.

2. Examine the slab base for breakage. If there is major damage, that section of the slab should be removed with the sledgehammer and replaced with fresh concrete. Before placing the new concrete, track down and correct the cause of the problem. For more details on replacing damaged concrete refer to chapter 10.

3. If the damage or sinking of the slab is relatively minor, use sand mix to bring the area back to the level of the surrounding concrete. Be sure to clean the original slab of any dirt, dust, soil, and organic matter before placing the sand mix. After it is placed, float it and allow it to cure for three to four days before replacing the paving.

4. To replace the paving, chip all the old mortar from the pavers with a hammer and chisel. Brush the bed and head joints clean of any loose flakes or dust.

5. Lay in a 1/2" mortar bed on the concrete slab.

6. Begin laying the pavers from an outside corner of the damaged area using the same size joints as in the original paving. Butter the head joints of each paver before placing it. Replace any broken pavers.

7. Strike the tops of the joints flat to prevent water entry. After the mortar is dry enough to be brushed without smearing, brush away any loose particles of mortar.

Chapter 17

WORKING WITH STONE

Stone masonry is one of the most lasting and beautiful types of masonry work. Stone has been used as a building and paving material for thousands of years, and its popularity shows no sign of waning. While stone masonry takes time and patience to master, many people prefer its free-form technique; you don't have to follow the exacting bond patterns found with brickwork. And since the stones don't have to be laid perfectly plumb and level, you have a lot of leeway in choosing the look you want.

TYPES OF STONE

There are two general classifications of stone: fieldstone, or rubble, and ashlar. As its name suggests, fieldstone is the common stone found in fields and along rivers and streams. Fieldstone can be either rough rubble or roughly squared rubble. Rough rubble is the crudest form of stonework. It utilizes irregular stones of different shapes and sizes; you use them in their natural, uncut state and make no attempt to lay them in courses. Since there's little or no dressing of the stones with a hammer and chisel, and because the mortar joints are simply raked out and brushed, rough rubble offers a very natural, rustic look.

Roughly squared rubble uses stones that are cut on the edges and laid in more or less level courses. While there's no specific repeating bond pattern, the stones should overlap those beneath them with few vertical joints in line. Roughly squared rubble thus has a more orderly appearance to it than does rough rubble.

Ashlar is quarried stone that's cut from the mountainside and squared on all edges. Lime-

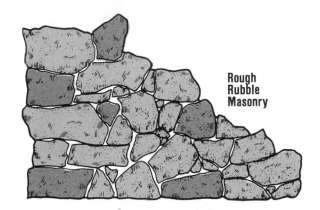

Rough Rubble Masonry

Roughly Squared Rubble Masonry

Variations of fieldstone

stone, sandstone, marble, flagstone, and slate are all examples of ashlar. This type of stone gives a neat but still irregular look. You can use a repeating bond pattern, or else vary it by laying two or three stones against a larger one.

Obtaining Stone

Where you get your stone depends primarily on the type you're looking for. If you're willing to invest a little time and effort, you might be able to get what you need for free. Good sources of building stone include abandoned quarries,

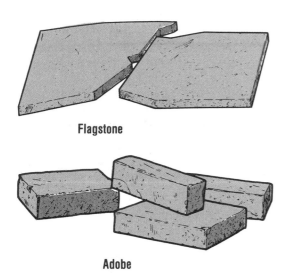

Flagstone

Adobe

Flagstone and adobe are two popular paving stones.

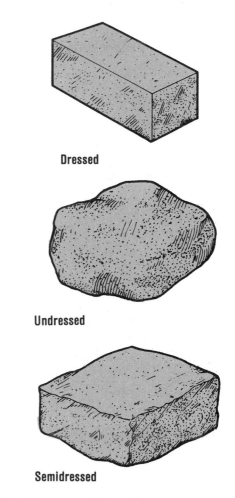

Dressed

Undressed

Semidressed

Grades of commercially sold stone

rock outcroppings in fields and mountains, stream beds and banks, old fence rows, and the crumbling foundation walls of deserted buildings. Building demolition contractors and road excavation crews are also good sources. Naturally, the larger the stone, the quicker the work goes. Large stones fill most of the space; you can use smaller ones to fill in between and conserve mortar. Avoid stones that you can't lift comfortably by yourself. Comfortably means able to lift for a sustained period of time.

Commercial quarries and stone yards sell stone in three grades: dressed, semidressed, and undressed. Undressed stones are uncut, unfinished, and generally the cheapest. Semidressed stone has squared-off corners, and the edges are roughly parallel. The stone isn't cut to any particular size. Dressed stone is cut to a specific size and sometimes can be ordered custom cut. This grade is laid in courses much like building brick or concrete block, and is the most expensive.

Most stone is sold by the ton, but some dealers sell it by the cubic yard. Required tonnage differs from undressed to dressed stone. One ton of rubble or undressed will generally cover from 25 to 45 square feet of wall, with an average thickness of 1'. One ton of dressed stone will cover approximately 50 to 60 square feet of wall, with an average thickness of 6".

Most stone yard or quarry personnel can estimate how much stone is needed for a given project. Calculate the cubic footage of the area to be covered and take your figures with you. Stone should match in color and texture with a good mix of sizes. If you're ordering dressed stone, specify the thickness you want and designate minimum and maximum lengths. Larger stones can probably be custom cut, if necessary.

TOOLS FOR WORKING WITH STONE

Many of the same tools needed for brickwork are used when working with stone. In addition, you'll need a sledgehammer with an ax-tapered head for breaking large stones and knocking off edges. A square-head mash hammer or ball peen hammer should be used for

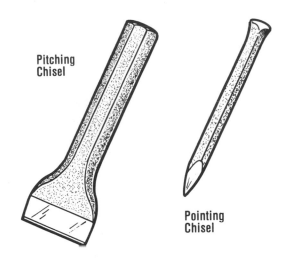

Pitching Chisel

Pointing Chisel

Stonemason's chisels

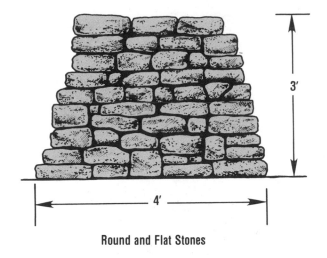

Round and Flat Stones

3'

4'

striking the chisel when making a cut; it's also ideal for settling stones in the mortar bed. Don't try to make do with an ordinary hammer. It's too light for stonework, and the face may chip from repeated blows on the chisel.

Stonemason's chisels are vital tools for cutting stone. The pitching chisel is much heavier and broader than the standard brick set and is used for facing stone, as in cutting a rock-faced edge. Trying to cut stone with a brick set will only ruin the blade. Instead, look for the grain and work along it with a hammer and pitching chisel. The pointing chisel is used to cut off bumps and rounded areas in order to make the face of the stone straight. Since all of the pressure of the point chisel is applied to a very small area, you can remove a bump or knob without cracking the rest of the stone.

What you use to tool the mortar joints greatly affects the appearance of the finished wall. A popular method is to smooth out the joint with a pointing trowel for a flush or full joint. A so-called rolling bead joint, in which the mortar is rolled out in a bead or convex shape, is formed with a concave stone beading tool. A rubblestone wall looks good with a raked out joint that can be formed with a wooden dowel or a piece of a broom handle. In all cases, wait to brush out the joints until after the mortar is dry enough not to smear. Ultimately, the choice will depend on your own personal tastes.

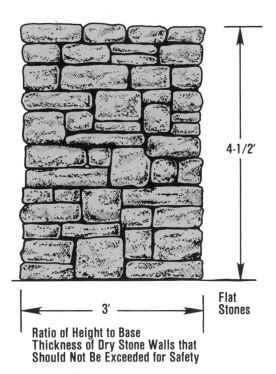

4-1/2'

3'

Flat Stones

**Ratio of Height to Base
Thickness of Dry Stone Walls that
Should Not Be Exceeded for Safety**

The safe height of a dry stone wall depends on the shape of the stones used.

A sturdy wheelbarrow is a must for transporting heavy stones. To move a stone, lay the wheelbarrow on its side. Wearing work gloves, roll the stone into the wheelbarrow; be careful not to twist your back. Work with a helper, if possible. Use both hands to push the wheelbarrow upright, gripping it by the sides. To avoid back injury, don't use the handles to reposition the wheelbarrow. And keep in mind that protective eyewear is essential when doing stonework because of the many chips that will fly during cutting.

DRY STONE WALLS

A dry stone wall uses no mortar, thus making it easy and inexpensive to build. It requires no foundation and can rise and fall from the effects of frost heave without cracking. The best stones to use for a dry wall are flat ones; the flatter the stones, the straighter you can hold the sides of the wall. Unless the stones you're using are of approximately equal size and fairly regular in shape, you'll need a good assortment of small ones to fill the spaces in between.

Because no mortar is used with this wall, the weight of the stones alone must hold the structure together. This means that you can't get carried away with the height of a dry wall. If the stones you're using are nice and flat, the wall can safely be one and a half times higher than its base width. Thus, if the wall is 3' wide, it should be no more than 4-1/2' high. It's important that you always use the base width for such measurements, since a stone wall often gets narrower with each succeeding course.

With moderately flat stones, keep the height and width equal. But if none of your stones are flat, or if you're using a mixture of flat and round stones, the height of the wall should be only three-fourths the width. In such instances, you need a 4'-wide base for a 3'-high wall.

Use the following procedure to construct a dry stone wall:

Laying the first course of a dry stone wall

1. Excavate to a depth of approximately 1' to form a solid bed. If desired, you can spread and tamp a layer of sand or stone for an extra firm base.
2. Using the heaviest stones for the first course, begin by placing a single row in the bed that marks the front of the wall. Adjust each stone so that its flattest side is on top and its next flattest side is facing out.
3. Use small stones and chips where needed to hold the large ones upright and stationary.
4. Lay a second row of stones in the same manner along the rear of the wall. Be careful not to place the large stones in direct alignment with those in the front row; instead, stagger them slightly for better stability.
5. Fill the space between the two rows with small stones and chips. Make them as level with the tops of the large stones as you can.
6. The first and second courses must be staggered. To do this, position the stones in the second course above the spaces between stones in the lower course. Again, use small stones and chips to keep the tops of the stones level.
7. Once you've positioned both rows of the second course, fill in between with small stones and chips. Two courses of the wall are now complete.
8. To lock the front rows to the rear rows, lay a third course of stones the entire length of the wall. These stones should be longer and as close to perfectly flat as you can find; each one should overlap as

much of the front and rear stones as possible.

9. Lay two rows of smaller stones to complete the third course, one on either side of the large ones.

10. Repeat this three-course pattern as many times as is needed to complete the wall.

11. You can cap the wall with a course of stones that stand out in some way, such as particularly large ones or those that are shaped differently from the rest of the wall. Or, if you prefer a perfectly flat top, apply a 2" layer of mortar over the top course.

MORTAR STONE WALLS

Mortar stone walls are not at all flexible, so they must have a base below the frost line to keep from cracking in freezing temperatures. Unlike dry walls, they can have perfectly vertical sides regardless of the type or size stones that are used. Mortar stone walls are also incredibly strong; nothing short of a sledgehammer can break them. The procedure for working with mortar and stone is the same no matter what type of stone you choose, whether it is irregular fieldstone or carefully cut ashlar.

Preparing the Mortar and Stone

To make mortar for stonework, mix one part cement with three parts sand. Use a sand that's somewhat close to the color of the stone. If the stone you're working with is dark, don't use a light sand unless you want a whitish mortar. Mix the cement and sand dry, then add water. When working with stone, it's best to keep the mortar a bit on the dry side; it must support the stones without sinking. As always, don't let the mortar sit for too long without using it, and mix only as much as you can work with in about an hour.

Since the stone used for this type of wall is often purchased (as opposed to a dry wall, which is usually made with "second-hand" stone that's very cheap, if not free), it's important to be able to come up with a rough estimate of the amount of stone you'll need. Probably the simplest way is to figure out the amount of cubic feet in the wall, then find out how much 1 cubic foot of the stone you're using weighs. Now you're left with a simple multiplication problem to arrive at the answer. As an example, if your wall will be 50 cubic feet, and 1 cubic foot of your particular stone weighs 10 pounds, then 50 cubic feet of your stone is 50 × 10, or 500; this means you need approximately 500 pounds of stone to build the wall.

Scrub the stones thoroughly with soap and water. Any dirt or other foreign matter left on them will affect the mortar bond. After cleaning the stones, let them dry in the sun. If you won't be using them for awhile, be sure to cover them to protect them from the elements. Applying mortar to wet stone will cause the mortar to lose much of its adhesive properties and make your job extremely difficult.

Building the Wall

After excavating and pouring the footer, proceed as follows:

1. Spread a 1" layer of mortar on the footer. Set the first stone in place; it's generally a good idea to use the largest and heaviest stones in the bottom course for better stability.

2. Butter the end of a second stone and place it alongside the end of the first

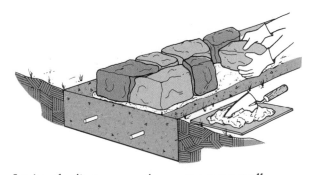

Laying the first course of a mortar stone wall

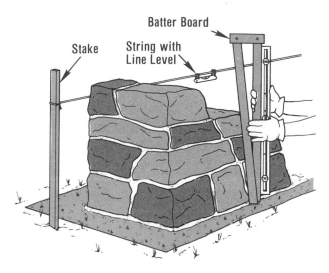

Check the wall for plumb and level.

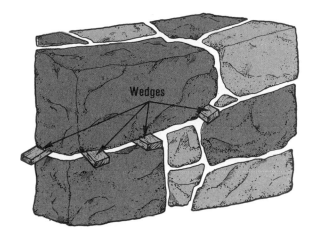

Use wedges to support large stones until the mortar sets.

one. Continue until all the stones making up the first course have been laid. Fill the spaces between stones with small stones, chips, and mortar.

3. Cover the first course with a 1"-thick mortar bed. As you lay the second course, make sure that the vertical joints of the two courses don't line up.

4. For all subsequent courses, lay a mortar bed and set the stones in place. Check that they are plumb and level as you go. You can build a batter board out of scrap wood to keep the wall plumb. If you don't have enough small stones and chips, mix gravel or crushed stone with sand and cement and use it to fill openings.

5. Because particularly large stones can squeeze the mortar out of the joints, it's a good idea to support them with small wooden wedges. These can be easily removed after the mortar has hardened, and the holes filled with fresh mortar.

6. Many people prefer deeply raked mortar joints in stone walls. To achieve this effect, use a piece of wood to rake out the joints to a depth of 1/2" to 3/4".

7. Top off the wall with a smooth, slightly pitched layer of mortar, or with pieces of slate or flagstone.

MORTAR ASHLAR WALLS

Ashlar, or cut stone, is easier to work with than rubble because the shapes are much more even. After excavating for the wall, proceed as follows:

1. If the wall is being built into a slope, step the footer as shown in order to save on concrete.

2. Depending on how far below grade the footer must be sunk to be underneath the frost line, stone can be saved by using concrete blocks to bring the foundation up to the grade level. Fill the hollows in the blocks with stone chips and mortar to make a firm base for laying the ashlar.

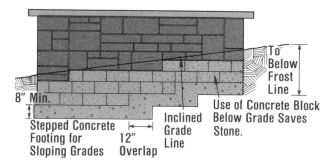

Step the footer when building into a slope.

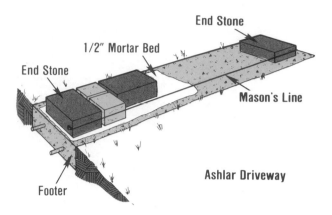

Lay the two end stones first.

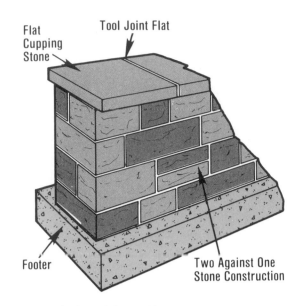

Topping off the ashlar wall

3. Cure the footer for three to four days before constructing the wall. Set a large stone, approximately as wide as the wall, at each end of the footer. Attach line blocks to the outside corners of these stones and run the mason's line between them to keep wall alignment.

4. Lay up the wall in two wythes, working from the ends toward the middle. Dry-lay stones first to obtain proper placement; remember to leave 1/2" head joints between the stones.

5. Lay a 1/2" mortar bed along the footer prior to building the wall. Mix only as much mortar as will be used in about an hour. Build from the ends toward the center, buttering the head joints before setting the ashlar in place. On wider walls the cavity between the wythes is filled with stone chips and mortar as the wall is built.

6. Depending on whether the stone is being laid in a random or coursed pattern, lay up the stonework in successive tiers, or courses, to the full height of the wall. Work both wythes simultaneously, filling the cavity as you go.

7. Lay end stones on each row first so that the line can be raised with the wall to keep it straight. The stone is plumbed "bump to bump" because the faces are not smooth like brick.

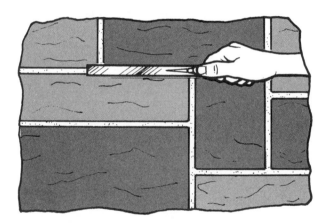

Raking out the mortar joints

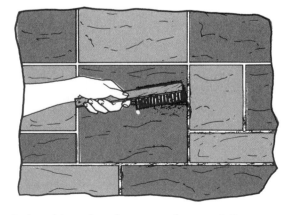

Begin brushing after the mortar has partially set.

8. Lay two smaller stones against a larger one (called a *two against one*) at intervals to strengthen the wall.

9. Lay large, thin stones on top as capping. The joints should be tooled flat to prevent moisture entry.

10. Rake out the joints to a depth of 1/2" after the mortar has set slightly to highlight the stone edges. Use a slicker jointer to do this.

11. Brush out the joints after the mortar is dry enough not to smear.

STONE RETAINING WALLS

The tools and materials needed to build this retaining wall are similar to those used for basic stone wall construction, with these additions: 4"-diameter drainpipe and a screen for covering the pipe. Don't forget to check your local ordinances and building codes before starting construction; however, stone retaining walls are rarely built higher than 4' or 5'.

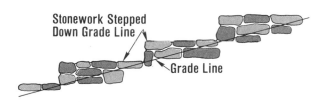

Stone retaining wall for sloped grade

1. As you build up the courses, note that rebar and other types of additional reinforcement aren't necessary on the stone wall because it's built lower to the ground than brick and block walls.

2. Because weep holes are essential for proper drainage, a length of plastic pipe must be installed approximately every 6' along the wall. Slope the pipe downward toward the front of the wall, about 8" above the front grade; this will help prevent pressure buildup against the back of the wall.

3. Place a screen over the back opening. Pile broken stone over the screen to prevent the pipe from becoming clogged.

To build a stone retaining wall for a sloped grade, follow the same procedure but construct the wall in steps to conform to the grade line.

STONE VENEER WALLS

When laying stone veneer over a concrete block wall, the procedure is very similar to the laying of a brick veneer wall. The procedure is as follows:

1. With masonry nails or a stud gun, attach wall ties to the wall every 2 or 3 square feet. (If the concrete block wall is being built from scratch, insert the ties in the mortar joints between blocks.)

2. Attach the stones to each other and to the wall with mortar. Lay a 1" mortar bed on the footer, and then begin setting

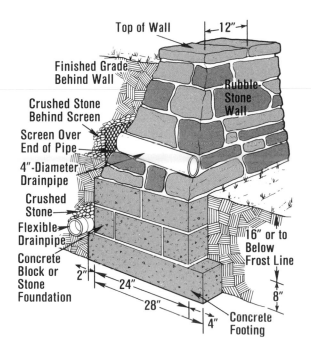

Stone retaining wall construction detail

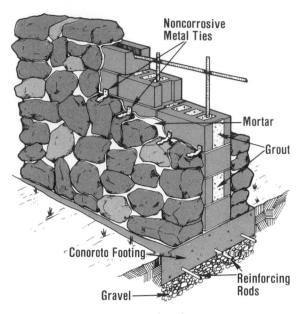

Noncorrosive Metal Ties

Mortar

Grout

Conoroto Footing

Gravel

Reinforcing Rods

Stone veneer construction detail

Stone veneer can be applied over rigid foam insulation.

the first course of stones. Unless you're using a high grade of dressed stone, you'll have to set small stones or gravel in the mortar to fill the irregular, open spaces where the large stones don't meet.

3. For each new course, build up a mortar bed and set the stones in place, checking the alignment as you go. Bend as many of the ties as possible into the joints between the stones.

4. Because very large stones can squeeze out all the mortar in the joints, support them temporarily with wooden wedges. When the mortar has set, pull out the wedges and fill the holes with mortar.

5. When a section has been laid, use a piece of wood to rake out the joints to a depth of 1/2" to 3/4". This will enhance the play of light and shadow on the face of the veneer.

6. To increase the insulating value of the veneered wall, a layer of rigid insulation board can be installed on the wall prior to placing the stone. Simply impale the foam board onto metal ties, pressing it firmly against the wall.

MANUFACTURED STONE VENEERS

Manufactured building stone veneers offer excellent durability and the authentic beauty of natural stone at a fraction of the cost of natural stone. The stone is made from portland cement, lightweight aggregates, and mineral oxide pigments. Its light weight permits easy installation over nearly any existing or new interior or exterior surface. Since the veneer creates a minimal

Mortaring the stone

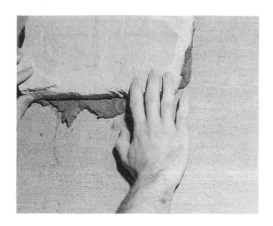

Applying the stone

shearing force, no wall ties, footings, or special structural supports are needed.

The veneer is set in a bed of standard type "S" mortar that is also used to grout the joints between the stone. The mortar should be mixed with enough water and lime to achieve a smooth, slightly sticky mix. Some manufacturers do not recommend the use of premixed mortars with their veneers.

On clean (unpainted, untreated, and unsealed) brick, block, concrete, or other masonry surfaces, the stone veneer can be installed directly on the surface. Extremely smooth concrete finishes should be roughened by etching with a muriatic acid solution. (See chapter 10.) On wood and other nonmasonry surfaces, metal lath and a scratch coat of mortar is applied prior to the veneer. The lath should be backed with building felt on outdoor applications. The preparation is the same as that used in stucco work (chapter 9). The scratch coat must be allowed to set before installing the stone veneer.

The following is a typical installation procedure for lightweight manufactured stone veneers:

1. Lay out the stone near the work area to give yourself a better choice of stone sizes and shapes.
2. In warm weather, dampen the surface of the masonry or scratch the coat before applying the veneer.
3. Select a stone and apply a 1/2" thick even layer of mortar to the back
4. Press the stone firmly into place on the wall surface so that the mortar behind the stone squeezes out around all sides. Using a gentle wiggling action while applying the stone will ensure a good bond. To help keep the stone clean during construction, start installing stone at the top of the wall surface and work down. Install the corner stones first for easiest fitting. When selecting stones, try to achieve a balanced pattern of shapes, sizes, colors, thicknesses, and textures. Keep the mortar joints between the stones as tight and uniform as possible. And avoid long straight unbroken joint lines.

Grouting the joints

Striking the joints

Brushing

Completion

5. When necessary, stone can be cut and shaped with a hatchet, brick trowel, or nippers to form special sizes and shapes for better fitting. Always try to position the trimmed stones on the wall surface so the cut edges will not show.

6. After all of the stone has been applied to the wall surface, fill a grout bag with mortar, and as if decorating a cake, partially fill the joints between the stones with mortar. (The stone supplier can furnish you with a grout bag.) Be sure while grouting, to cover any noticeable broken stone edges with mortar. If you prefer colored mortar joints, oxide colors can be mixed with your mortar prior to grouting.

7. When the mortar joints become firm, use a wooden or metal striking tool to rake out the excess mortar to the desired depth and at the same time to force the mortar into the joints to thoroughly seal the joint edges. Be careful not to work the joints too soon or the mortar will smear.

8. Brush the mortar joints with a whisk broom to smooth them and clean away the loose mortar. At the same time, brush off any mortar spots from the face of the stone. Loose mortar and mortar spots that have set for only a few hours clean up easily and should never be allowed to set up overnight.

9. The use of a high-quality waterproofing sealer is highly recommended, especially on surfaces exposed to severe freezing and thawing, excessive moisture, or conditions that could discolor or stain the stone. A sealed surface is much easier to clean than an unsealed surface.

STONE BARBECUE

Stone's sense of mass might seem too heavy and imposing for a simple barbecue, but its natural qualities work quite well to create an informal, casual air for garden cookouts. This barbecue is a relatively easy and economical project within the abilities of even the novice do-it-yourselfer. It uses firebrick, fieldstone, and flagstone to achieve a unique look. The size of the barbecue will be determined by the dimensions of the grill you choose to use.

1. Excavate an area 2' wider and 1' longer than the grill, cutting the edges as straight as possible. Dig to a depth of about 4".

2. Pour concrete mix into the hole and screed it smooth, sloping it slightly away from the end where the fire will be built. Fill to about 1/2″ below ground level. Cure the concrete for three to four days.

3. Center the grill on the edge of the slab away from where the fire will be built. Mark the edges of the grill on the slab, then remove the grill and lay a 1/2″ mortar bed within the marks.

4. Lay a bed of firebrick in the mortar, taking care to keep the bricks straight and to maintain a slight slope toward the edge of the slab. Joints between the bricks should not exceed 3/8″. Sweep dry mortar mix into the joints to fill them.

5. Construct the firebrick lining beginning at the outer edges. Build up the brick in three courses with mortared joints. Use a level to keep the liner straight and plumb.

6. Construct the stone shell around the liner by setting the fieldstone in a mortar foundation not thicker than 2″. Dry-fit the stones before setting them into the mortar. The shell is constructed outward from the liner and built up in courses.

7. After the stones in each course are set in the mortar bed, add mortar between them to fill the cracks.

8. Construct the sides of the shell 2″ to 3″ higher than the liner; construct the end 3″ to 4″ higher than the sides.

9. Spread a mortar bed on top of the shell and lay the flagstone. To prevent moisture entry, strike the joints between the stones flat. It's also a good idea to lay the flagstone with a 1″ overhang to act as a drip edge.

FLAGSTONE WALKS

Constructing a flagstone walk is a simple way to improve a lawn or garden. Because of its

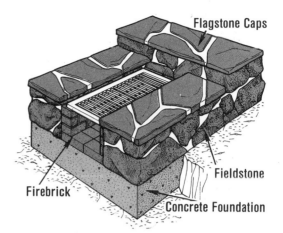

This durable stone barbecue will provide years of service.

size and weight, flagstone can be laid directly on a stable, level soil base. The 2″-thick variety should be used in the following procedure in order to prevent cracking; flagstone of lesser thickness may not hold up under traffic.

1. Select stones whose length is equal to the desired width of the walk. Because 2′-long flagstone is common, this example provides directions for laying a 2′-wide walk.

2. Mark off an approximately 2-1/2′-wide path for the walk, and cut the grass very short along the entire path. Sweep it clean of all debris.

3. Beginning at one end, lay the flagstone down one after another, with a small space between each one. You can use a length of 1×2 wood to help space the stones evenly.

4. If you reach the end of the walk and discover that the remaining space is too small for a single stone, you have two choices. Either readjust the spaces between the stones accordingly, or cut a stone to fit.

5. For a perfectly curved path, drive a stake into the ground and tie one end of a string to it. Swing the other end of the string in an arc around the stake, driving an additional stake every few feet as

you go. Connect the stakes with string, then lay the stones along the string guide to form the curve.

6. When all of the stones are in final position, throw loose soil around them, followed by grass seed. Tamp the soil firm.

Cutting Flagstone

If you find it necessary to cut flagstone, you can do it easily with a power saw and abrasive blade. If you don't have access to such tools, you can make the cuts with a 3"-wide stonemason's chisel and a small sledgehammer as follows:

1. Mark the cut line on the stone with a piece of chalk.
2. Lay the stone over a length of pipe or a 2×4, with the cut line directly over the pipe or wood.
3. Cut a groove in the stone by striking it lightly with the hammer and chisel along the cut one. Then turn the stone over and repeat the process on the other side.
4. When both sides have been grooved, move the stone so that the cut line is about an inch beyond the pipe or wood. Using just the hammer, strike the stone until it snaps at the cut line.

FIELDSTONE WALKS

To use fieldstone for a walkway, the individual stones must have at least one flat side. Fieldstone walks are a bit more work than flagstone walks, but you can still do one easily in an afternoon.

1. As with a flagstone walk, begin by marking off an approximately 2-1/2'-wide path, cut the grass along the path as short as possible, and sweep it clean.
2. Starting at one end, set a stone in place, flat side down. Using a knife or the point of a trowel, trace the outline of the stone in the ground.

3. Remove the stone. Now dig a hole within the outline that's roughly the size of the rounded side of the stone, but make it about 2" shallower.
4. Place the stone, round side down, into the hole. If you dug the hole correctly, the level face of the stone should project about 2" above the ground.
5. If the stone is too low, simply remove it and add some dirt. If it's too high, dig the hole a bit deeper.
6. Place the rest of the stones in a similar manner, leaving 1" to 2" between the stones.
7. When all of the stones are in place, retrieve the pieces of sod you removed when digging the holes. Fit them, grass side up, around the stones and tamp them down firmly.

With this type of walk, the stones have a tendency to gradually sink into the ground over time. When this happens, just pick them up, spread sand or soil in the hole, and lay them back in place. With regular maintenance, your fieldstone walk will last for years.

STONE STEPS

The first thing you should do when building stone steps is figure out the rise; this is the vertical distance from the top of the stairs to the bottom. To accomplish this, drive a tall stake into the ground at the base of the incline. Then string a line horizontally from the top of the incline to the stake. The rise is measured from the point where the string meets the stake to the ground. Divide the total rise into any number of equal parts to establish the rise of each step. For comfortable climbing, a step rise should be no higher than 8". This project uses a 6" rise.

Good measurements for the tread (the part of the step you walk on) are a width of 24" and a depth of 12". But because part of the treads will be covered by stones supporting the risers and side walls, add about 4" to the tread depth and 6"

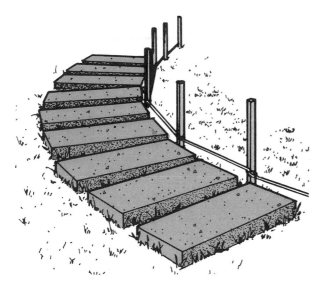

Use string and stakes to lay a curved walk.

to the width. So the treads for this project will be 30″ wide and 16″ deep. Stone for treads should be 2″ thick. Flagstone is excellent for step treads as is fieldstone—provided you can find enough pieces that are the right size, shape, and thickness. Never use slate; it becomes very slippery when wet.

You'll also need a good number of medium-sized stones to support the vertical face of risers and side walls. Pieces should be roughly 4″ high and 4″ wide. Lengths can vary. Proceed as follows:

1. Cut into the earth to form a step 4″ high, 16″ deep, and 30″ wide. The rear of the step tread surface should be slightly higher than the front. Make the riser face as vertical as possible.

2. Place a row of roughly 4″-high stones tightly against the riser face. If any stones sit too high, push them deeper into the ground. Shim up low stones with additional dirt fill beneath them.

3. Set one of the stone treads on top of the step, with the front edge extending about 1″ beyond the row of riser stones. Use a level to make sure that the back edge of the tread is slightly higher than the front to prevent standing water. If

necessary, lift the tread and throw some soil under it to provide an incline.

4. Cut another step 4″ above the top of the first tread. Use the same 16×30 dimensions for all the steps, and always make the rear a bit higher than the front.

5. Place a row of 4″-high stones across the rear of the first tread and tight against the second riser. Lay the second tread in place and check it with the level as before. Repeat this technique until all of the steps have been installed.

6. To hold the earth back and the stones in place, build a small retaining side wall on each side of the steps. Use the same procedure outlined earlier in this chapter, keeping in mind that the entire width of the walls does not have to rest on the treads.

FLAGSTONE PATIOS

Flagstone is a favorite patio paving material. From an informal staggered pattern with grass sprouting out of the joints to a tightly laid arrangement using precision-cut stone, you have a wide range of looks from which to choose. The foundation for a flagstone patio can be done several ways, depending largely upon the climate where you live and the type of soil you have to work with. In a cold climate with harsh winters where the soil is subjected to heaving from freeze–thaw cycles, a reinforced concrete base must be used. If your soil is a heavy clay type, use a thick, porous subgrade. This project has a reinforced concrete base, a crushed stone subgrade, and a mortared flagstone surface.

1. To begin, complete the basic patio excavation. Dig to the appropriate depth for the total thicknesses of the foundation layers to be used.

2. Install the forms and place stakes at 3′ intervals for support. For additional bracing, you may want to place boards against the outside of the forms.

Lay out the flagstone in a dry run before mortaring.

Spread mortar generously before laying a stone.

Making sure the stones are level

3. Install screed guides across the width of the excavation to keep the screed board level when you smooth out the freshly poured concrete. Nail the guides to stakes, keeping in mind that they'll eventually be removed.
4. Cover the excavation with 2″ to 4″ of crushed stone. To make sure the stones are level, lay the screed board over the guides to make a visual check.
5. Place wire mesh reinforcement over the stone. The mesh may have to be placed on support blocks to keep it in the middle of the concrete during pouring.
6. Pour the concrete, then use a long 2×4 screed board to smooth and level it. With a helper holding the other end, make a slow, careful sawing motion across the guides. The slab surface must follow the slope of the excavation.
7. When the entire surface is level, carefully remove the screed guides and stakes, then fill in the gaps with additional concrete.
8. Smooth over the entire surface with a bull float to eliminate footprints and lumps caused when removing the guides.
9. Let the smoothed surface partially dry overnight, then scrape it with a rake in a crosshatch pattern; a textured surface will be more effective in holding the stones in place.

10. Wait about two days, until the concrete has dried fully, to lay out the stone pattern. With irregular stones, try to stay with the natural shape and size as much as possible. You may have to go through a few dry runs before getting the look you want.
11. To make a 3-1/2″ wide, 1-1/2″ thick mortar border around the edge of the patio, extend the height of the existing forms by nailing 1-1/2″ thick wood strips directly to the tops of the forms. Place 2×4s on the surface of the concrete base and against the extended forms. The top of the 2×4s should be flush with the top of the extended forms. Butt the flagstone against these 2×4 guide boards. After the stone has been set in mortar, remove the 2×4 guide boards and fill in the area with mortar.

Smooth the mortar with the face of the trowel.

Washing off the flagstone surface

12. Working with a few stones at a time, spread a layer of mortar under each stone. If any stones aren't level, either tamp them down or place additional mortar under them. Then trowel additional mortar into the joints between stones and smooth out the surface, scraping the excess mortar off the stones.

13. Let the finished surface set for a few weeks, then remove any mortar stains from the stones by cleaning with a weak acid solution of one part muriatic acid in twelve parts water. See chapter 10 for details. Flush with plenty of clean water to neutralize the acid.

If you prefer a casual, irregular appearance for your flagstone patio involving little excavation and no mortar, you can use a soil base. Just set the stones directly on the lawn and score around the edges with a shovel. Remove the stones and dig a shallow hole for each one, allowing room for sand as a subgrade. The sand will seat the stones more firmly in the soil, preventing rocking and eventual cracking. If a raised area is desired, don't excavate under the stones. Instead, remove only the grass, and lay the stones directly on stop of the existing grade. Fill in the joints with a soft paving material or with topsoil to grow lawn or groundcover for a natural look. You can follow this method for laying a fieldstone patio as well. In fact, you might feel that the irregular shape of fieldstone lends itself better to a casual look than does flagstone.

MAKING REPAIRS TO STONE

While it's certainly true that stone masonry is one of the most lasting and durable forms of construction, some type of repair work is bound to be needed sooner or later. Following are some common repair techniques specific to stone that you can follow to keep your work looking good. Always wear safety goggles and work gloves when doing any type of stone repair.

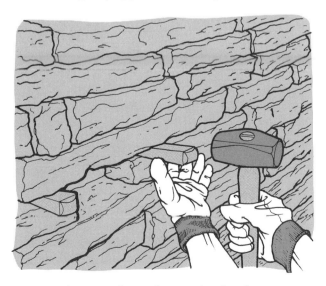

Using wedges to enlarge the opening for the stone

Replacing a Dry-Laid Wall Stone

A dry stone wall can become loose over time, and some of the stones can fall out. You may be able to simply push a fallen stone back in place. Force it into the hole; if necessary, tap the stone with a sledgehammer, but be careful to cushion the blows with a wood block.

If the wall has settled to the point where the stone can't be fit back in place, you'll have to lift up the stones above the opening with stone wedges. Using a small sledgehammer, drive a wedge between each stone and the stone below it. Drive in as many wedges as needed to enlarge the opening and get the stone back in position. When the stone has been replaced, pull out all the wedges.

Refitting a Dry-Laid Cap Stone

The cap stones of dry walls are especially prone to shifting and loosening due to frost heave, so it's a good idea to inspect them periodically. You can use a pry bar to reposition a cap stone that's out of place. If it can't be repositioned securely, permanently wedge a few small stones under it to level it. You can also try lifting the stone out of the wall entirely and dropping it back into place from a height of about 6" to reposition it.

Replacing a Mortared Wall Stone

If a stone in a mortared wall becomes loose or damaged, it must be replaced using the following procedure:

1. Using a cold chisel and ball peen hammer, cut away the joints around the loose stone. To remove as much of the mortar as possible, use a mortar hook to rake out the joints. Work carefully so that you don't damage any of the adjoining stones.

2. If the stone isn't loose, you may need a pry bar to work it out. Fit the bar in the joint below the stone and work it back and forth. Repeat on each side of the stone until it's loose enough to be removed.

3. With the stone removed, use the chisel and hammer to remove any remaining mortar from the hole. Use a stiff-bristled brush to wipe out the loose particles.

4. Wash both the hole and the replacement stone with clean water.

5. Spread a 3/4" layer of mortar on all sides of the hole. You may want to place small stones in the bed joint to keep the stone steady while the mortar sets.

6. After buttering the top and backside of the stone with mortar, push it straight into the hole, aligning its face with those of the adjoining stones.

7. Pack mortar into the joints using a trowel. Wipe off any excess mortar with a dampened piece of burlap or rough cloth.

8. After tooling the joints, use a fine spray mist to keep the mortar damp for about three days until it cures completely.

Repairing a Decorative Edge

Minor damage to a decorative stone edge can be rebuilt, but this procedure should be used

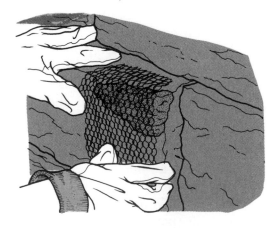

Use the wire mesh patch to shape the mortar.

only on structures that aren't load bearing, since it provides no structural support.

1. To collect the stone dust for use later, set a container below the damaged area while you work.

2. Use a cold chisel and ball peen hammer to chip away the edge until you reach solid stone. For good bonding of the mortar patch, chisel several 1/4" grooves into the damaged area.

3. After wiping away any loose particles with a stiff-bristled brush, wash the edge with clean water.

4. Use a putty knife to apply a thin, even layer of mortar to the edge. Let it set about fifteen minutes, then use the blade of the putty knife to scratch grooves in the mortar.

5. Apply additional mortar, building it up to form the new edge.

6. Cut a piece of stiff wire mesh the size of the edge for use as a mold. Position the mesh against an undamaged edge to duplicate the shape, then press the mesh into the mortar to shape it.

7. Remove the mesh and smooth out the mortar with the putty knife. You may have to wet the knife to work the mortar into the desired shape.

8. Take whatever stone dust you've collected and, before the mortar stiffens, sprinkle the dust over it. Don't put any chunks of stone in the mortar; crush them into powder with a hammer.

9. Keep the repair damp for about three days until it cures by taping a piece of plastic over the surface.

Chapter 18

FASTENERS FOR CONCRETE, MASONRY, AND STONE

As often shown in the various projects illustrated in this book, the finest method of attaching anchor bolts, hooks, brackets, and other hardware to concrete, brick, block, and stonework is during the construction process. Hardware can be embedded in fresh concrete; in the mortar joints between courses of brick, block, or stone; or in the grouted cores of hollow masonry

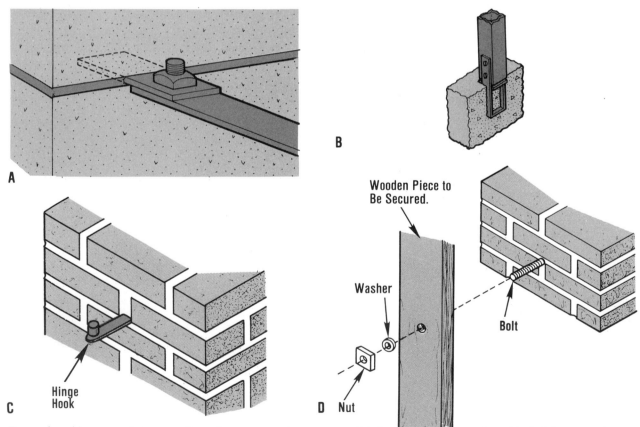

Examples of fasteners integrated into the construction process: (A) simple metal strap placed in brick mortar joints is used to attach decorative wrought iron, (B) metal U-bracket embedded in concrete pier foundation will attach a 4×4 wooden deck post, (C) hinge hooks for gate set in mortar joints, and (D) wood frame bolted to back post for gate attachment.

278

TABLE 18–1: FASTENERS FOR CONCRETE, MASONRY, AND STONE MATERIALS

Fastener Type	Application	Materials	Fasteners	Remarks
Masonry nails	Solid	Concrete, block, brick	—	Wooden furring most common application. Can be difficult to drive.
Fiber and neoprene plugs	Solid or hollow	Concrete, block, brick	Wood, sheet metal, or lag screws	All purpose anchor. Vibration-resistant.
Plastic and nylon anchors	Solid or hollow	Concrete, block, brick	Sheet metal screws	Economical, easy to install. Good for lightweight jobs.
Lead screw anchors	Solid or hollow	Concrete, block, brick	Wood, sheet metal, or lag screws	Not recommended for constant vibration or shock loads.
Lag screw anchors	Solid	Concrete, brick, mortar joints	Lag screws	Internally threaded. Use longer anchors in weak materials.
Machine bolt anchors (caulking)	Solid	Concrete, brick, stone	Machine screw or bolt	Can be set flush or recessed in surface. Design allows for slight hole irregularities.
Machine bolt anchors (noncaulking)	Solid	Concrete, brick, stone	Machine screw or bolt	Double units designed for heavy-duty application. Can be set flush or recessed. Double units not for use in stone.
One-piece stud anchors	Solid	Concrete, brick, stone	None	Tapered top protects threads. Some designs not for use in stone.
One-piece expansion bolts	Solid	Concrete, dense brick, stone	None	Truly permanent anchors.
Nail-in anchors (nylon body with screw)	Solid or hollow	Concrete, brick, block, plaster, wallboard	None	Fast installation for light-duty holding power. Removable screw.
Nail-in anchors (rivet design)	Solid	Concrete, brick, block, stone	None	Strong for size. Easy to install. Resists vibration.
Rivets	Solid	Concrete, brick	None	Very light work, such as signs.
Toggle bolts	Hollow	Block, hollow tile, stucco over lath, plaster over lath, wallboard	None	Distributes load. Good for weak materials. Installation can be cumbersome. Fastener ruined if removed.
Expansion (Molly) bolts	Hollow	Block, plaster, wallboard, hollow tile	None	Once set, screw can be removed without ruining fastener; must be properly sized.

units. Doing this makes the hardware an integral part of the structure and usually provides the greatest amount of holding power possible.

In many cases, the hardware set in place during construction provides a connecting point for additional structural units. Anchor bolts secure wooden sills to concrete or block foundations. Brackets set in concrete piers tie into the 4×4 posts used to support a wooden deck or carport. Hinge brackets set in the mortar joints of a brick pilaster provide mounting points for a gate.

Obviously, any hardware integrated into the concrete or masonry structure must be located and positioned with extreme care. Misaligned pieces set at odd angles to the mounting surface create more problems than they solve. So in any new construction project take the time to anticipate any fastening needs and install the right hardware during the job.

SELECTING FASTENERS

Unfortunately, not all future needs can be predicted. Sooner or later, everyone faces the task of attaching something to an existing concrete, masonry, or stone surface. How you solve the problem depends on a number of factors, the first of which is selecting the right type of fastener for the job (table 18–1, on page 279).

Fasteners can be divided into two major groups—those designed for use in solid materials, such as concrete, mortar, grout, brick, and stone, and those intended for use in hollow walls,

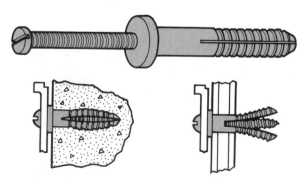

This expandable nylon anchor demonstrates the holding principle in both solid and hollow walls.

such as cored concrete block or stucco over wood sheathing.

Fasteners designed for use in solid materials actually anchor themselves against the inside of the hole into which they are placed. The shell or plug portion of these fasteners expand and grip the hole sides as the unit is tightened. Some of the fasteners used in solid walls can also be used in hollow wall applications.

Fasteners for hollow wall applications actually pass through the face of the wall to the hollow cavity behind it. They generate their holding power when part of the fastener expands and grips against the backside of the wall.

HOLDING POWER AND WORKING FORCE

Always keep in mind that the holding power of any fastener, solid or hollow design, will depend to a great extent on the strength and tightness of the material into which it is set. Table 18–2 lists the relative strengths of various types of concrete and masonry materials. Weaker base materials will require additional and/or larger fasteners than their stronger counterparts.

In addition to the strength of the wall material itself, you should also consider the conditions under which the fastener will be installed and used. Will it be necessary to remove the fastener from time to time, or is the mounting permanent? How far apart will the fasteners be spaced, and how deeply will they penetrate the wall or floor?

When determining the size and number of fasteners to use, the most important consideration is the weight of the object to be held. The weight, along with the type of loading, the angle of loading, and the way in which the load will be applied, determine the maximum working force of the fastener.

There are different forces that affect the holding power of the fastener to be used. *Shear* is the downward pull exerted on the fastener by the item it is holding. For example, a picture hanging from a nail that has been horizontally

TABLE 18–2:   CONCRETE AND MASONRY HOLDING STRENGTHS

Material	Strength	Remarks
Placed concrete	Excellent	Assuming a quality mix. Large aggregate mix may cause difficult drilling.
Mortar and grouts	Good to moderate	Low-quality mixes have poor holding power. Joint thickness limits anchor size.
Concrete block	Good	Locate hollow wall fasteners away from webs.
Brick	Good	Cored brick may limit the use of solid wall anchors. Power drill only to prevent splitting.
Stone	Moderate to Excellent	Sandstone and limestone are easiest to work. Drilling in granite, marble, and denser fieldstone may be difficult.
Stucco	Poor	Drill carefully through top, brown, and scratch coat layers. Penetrate into backing wall or wood studs to increase holding power.

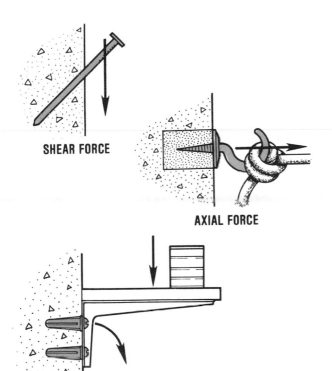

SHEAR FORCE

AXIAL FORCE

COMBINATION FORCE

Types of forces placed on fasteners

driven into the wall would constitute shear force. The main consideration would be whether or not the weight of the picture would bend the nail. This, of course, is true if you assume that the material into which you are driving the nail will not crumble.

An *axial force* exerts an outward pull on a fastener. An example of a fastener that must sustain this type of load is a fastener used to hang a clothesline.

A *combination force*, on the other hand, consists of an outward and a downward pull by the load. A bookshelf supported by a bracket exemplifies this type of force. It is necessary for the bracket not only to hold the load up, but to keep it from pulling out as well. A load of this kind requires a different fastener (kind and size) than the picture hanging on the wall.

The type of load can also vary. Static, or dead, loads do not change in weight. Other loads may consist of sudden impacts or vibrational forces. A good example of a vibrating load is the brackets and fasteners used to mount an air conditioner. Over time, the operational vibrations may cause some fasteners to loosen and pull out. A solution would be a screw-type fastener that could be periodically retightened when needed. Certain sizes of hardware and fasteners may do a good job supporting a wrought iron railing or gate, but what will happen the first time an energetic child does a little climbing or goes for a swing?

The fastener you select must be able to support the heaviest projected load, and do it with a reasonable margin of safety. The basic rule to keep in mind for providing a safety factor is: Never use a safety factor of less than 4 to 1 for general fastening, and whenever extreme vibratory conditions are involved, use a 10 to 1 ratio. In other words, if you want to hang something on the wall that weighs 25 pounds, use a fastener with a maximum holding power of 100 pounds. This will provide a safe and secure mount.

ANCHORING CEMENTS

As described in chapter 9, special anchoring cements are available for positioning hardware in a cut-out section of existing concrete (see page 144). Anchoring cements are designed to expand slightly as they cure to create a tight-fitting patch that holds the hardware securely.

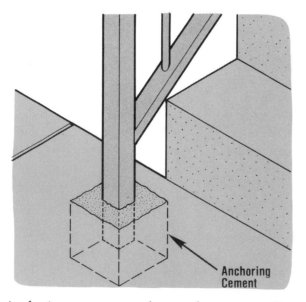

Anchoring cements are often used to secure railings and other larger fixtures to existing concrete surfaces.

Anchoring cements are ideal for mounting hardware that will be subject to high loads or vibrational forces. Anchoring cements are limited to use in solid concrete surfaces. The size and depth of the mounting hole depend on the size of the hardware being anchored, a factor that can limit some applications.

MASONRY NAILS

Masonry nails are specially hardened so they can be driven into concrete and masonry. Their most common applications are attaching wooden furring strips to walls or mounting plates to floors. Masonry nails are not designed to be used in a load-bearing capacity, such as in mounting brackets for shelves or other similar items.

The hardening process makes masonry nails brittle, so they may snap off if they are not struck squarely. A flying nail can be extremely dangerous, so always wear full safety goggles and keep the area free of spectators. A hard hat and heavy work gloves are also recommended.

Driving masonry nails requires a heavier-than-normal hammer, such as a hefty ball peen or light sledgehammer. Drive the nail straight in by hitting the nail squarely. If the nails seem particularly difficult to drive, consider using a fastener that requires a drilled hole.

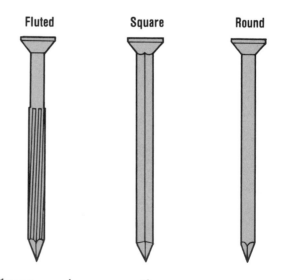

Three types of masonry nails

STUD FASTENERS

To simplify the task of driving nails into masonry, a *stud driver* can be used. This special tool is designed to set masonry nails or stud fasteners in concrete, bricks, and even soft metals. The driver tool prevents the nails from buckling and holds them upright until they are driven home.

Stud fasteners from 3/4" to 3" are usually available. When selecting the proper length, remember that the fasteners should be long enough to penetrate from 1/2" to 1" in masonry block and from 3/4" to 1-1/4" in mortar joints. Be sure to add the thickness of the material you are attaching to the amount of penetration needed for the best holding power. Most stud fasteners can hold loads up to 200 pounds each.

Various types of stud fasteners are available. The nail type is generally used to nail strips of lumber onto masonry surfaces. These can be driven directly through the lumber into the concrete. The threaded studs have a nail-like point that is driven into the masonry surface so that

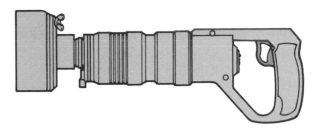

Power-actuated stud driver. These tools are extremely dangerous if misused.

the threaded end is left projecting. Nuts can then be set over the threaded end to hold the fixture in place. The washers that are attached to the fastener serve three purposes: they aid in positioning the stud in the stud driver; they act as a guide to the stud while it is being driven home; and they offer a better grip after the object has been fastened in place.

To use the tool, a stud fastener is inserted into the holding sleeve or tube of the driver. An anvil in the other end of this sleeve projects at the top so that it can be struck with a hammer to force the pointed portion of the stud fastener into the masonry. The base of the driver is held snugly against the surface. It is best to use a light sledgehammer for the striking tool, particularly if there are many studs to set.

If you are driving a great number of studs, such as in a basement remodeling project, it may pay to purchase or rent a power-actuated stud driver. One type of power driver uses .22 cartridge loads to fire the stud fasteners into the concrete or masonry. (Compressed air drivers are also available.) Misuse of these high-velocity stud drivers can be dangerous. Follow all manufacturer's safety precautions to the letter at all times.

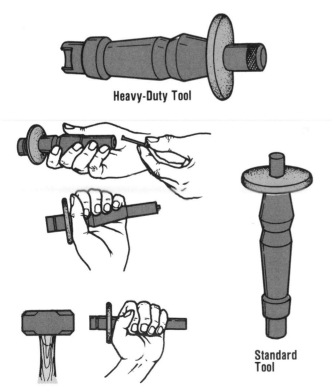

Heavy-Duty Tool

Standard Tool

Manual stud driver for fastening wooden furring strips to concrete and masonry walls

HOLES FOR FASTENERS

The successful use of most add-on masonry fasteners depends on a properly drilled hole. The importance of this cannot be overstressed. The diameter of the hole must be drilled to the exact size required by the particular fastener. This diameter is listed on the fastener package or may be stamped on the fastener itself. Even slightly oversized holes will drastically lower the fastener's holding power.

The availability and versatility of portable power drills, including cordless models, makes drilling by hand seem like an ancient art. But while a power tool should always be your first choice for precise, speedy drilling, there may be times when you need to work by hand.

Drilling by hand requires the use of star drills and a light sledgehammer. Be sure to wear the proper eye protection. Hold the drill perpendicular to the masonry surface and rotate it as you strike it with the sledgehammer. Work in stages to ensure precise results. If the fastener calls for a 3/4"-diameter hole, drill a 1/4"-diameter pilot hole first. Enlarge this to 1/2" and the final 3/4". As you can see, hand drilling can be quite a bit of work. Care and precision are needed to prevent oversized or angled holes. A series of light-to-moderate strikes always works better than several heavy blows.

The easiest way to achieve a proper hole in brick, concrete, mortar, or stone is to use a portable electric drill equipped with a carbide-tipped masonry drill bit. Powerful, slow-turning drilling provides the best results. Use a variable-speed drill with a maximum chuck diameter of 3/8" or 1/2". Smaller 1/4" chuck drills do not generate sufficient power for drilling in masonry.

Make sure the bit is loaded correctly and seated tightly in the chuck. If the hole must be a specific depth, wrap masking tape around the bit as a depth gauge. Be sure to wear the proper eye protection. Position the bit against the masonry

Power drilling with a carbide-tipped masonry bit

and squeeze the trigger slowly. Push straight in; bending forces will easily snap the brittle carbide-tipped bits. Maintain a steadily increasing pressure as the bit enters the work. Do not let the drill ease up or run idly in the hole. Once the hole is started, increase the pressure as the bit cuts away at the material.

Although coolants are not generally necessary, using water or turpentine as a coolant may be desirable when drilling hard material such as certain types of stone. When a coolant is used, the entire drill point should be kept wet.

Keeping the hole free of dust is important when using a masonry drill. A drill having a flute or twist along the body will itself remove the cutting dust. However, when excessive moisture is encountered, particularly in horizontal or downward drilling, it may be necessary to lift the drill slightly from time to time to clean the hole. If the hole is deep, blow it out or flush it out with water from a small syringe.

The drilling of small holes is never a problem, whether the material is hard or soft. Large diameter holes, however, should first be drilled to a small diameter and then to a larger diameter, depending upon the hardness of the material and the diameter of the desired hole. This also helps to avoid the tendency to make the hole oversized. It makes a neater, more precise hole. When drilling holes in concrete, examine the hole occasionally to avoid drilling through large pebbles. If a pebble is encountered, it is best to break it with a center punch and hammer to protect the drill point.

Forming a hole manually using a star drill and heavy hammer

FASTENERS FOR SOLID WALLS AND SURFACES

Fasteners used in solid walls include such items as screw anchors, machine bolt anchors, one-piece anchors, self-drilling anchors, and nail drive anchors. These are major categories. There are literally dozens of solid wall anchor designs available, but most use one of the holding principles outlined in the following sections. Remember, always follow the manufacturer's exact recommendations for use, hold capacity, and drilled hole size.

Screw Anchors

These fasteners are two-piece inserts consisting of a screw and some type of anchor. The anchor is set into the drilled hole. When the screw is turned into the anchor, it presses the anchor against the hole walls, binding it in place. The expandable anchor portion of the fastener can be any of several designs.

Fiber Plugs. Fiber plug anchors are designed for use with standard wood, sheet metal, and lag screws. Most plugs are formed of braided jute that is compressed into a tubular shape and chemically and heat treated to give it a tough, elastic consistency. A soft lead lining on the inside of the tubular plug allows the screw to cut its own thread. It also prevents the screw from cutting the jute fibers. Fiber plugs are manufactured in a complete range of sizes to match most screw sizes.

Installation is very simple. Drill the proper-size hole in the concrete or masonry. Install the fiber plug anchor in the hole, and position the bracket or hardware to be secured. Finally, turn the screw into the plug lining. As the screw is turned in, the flexibility of the jute allows it to be compressed into all irregularities in the walls of the drilled hole.

Since a fiber plug depends on a high level of compression for its holding power, it is absolutely essential that the plug, drill bit, and screw sizes are all the same. For example, when using a No.

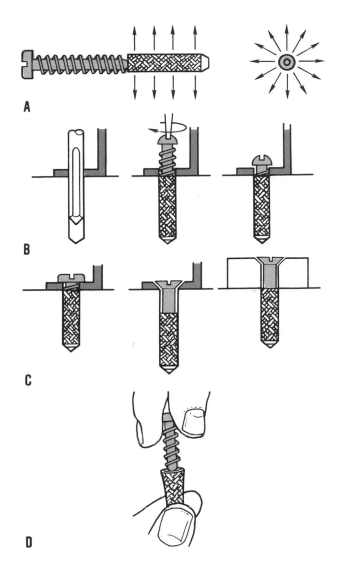

Fiber plug screw anchors: (A) holding principle, (B) basic installation, (C) examples of proper length plugs, and (D) flaring the end of the plug for use in hollow wall applications

12 screw, use a No. 12 fiber plug, and drill the hole to the diameter specified on the fiber plug packaging. This is usually the same diameter as the screw body, which in the case of a No. 12 screw is 1/4".

The shoulder of the screw should not enter the plug. Forcing the screw shoulder into the plug adds nothing to the fastener's holding power and may bind or shear off the screw head. Use a plug that is only as long as the threaded portion of the screw. If you are using sheet metal screws (threaded to the head), the plug can be the length of the screw minus the thickness of the hardware to be fastened. If the shoulder of the screw is longer than the thickness of the hardware to be fastened, countersink the fiber plug in the masonry deep enough so that the shoulder will not touch it.

Neoprene Sleeves. Anchors are also available made of neoprene, a synthetic rubber material often used for mounting window fans, stereo speakers, and other high-vibration items. They are installed and held in the same manner as the fiber plugs just discussed.

Both neoprene sleeves and fiber plugs depend on the strength of the wall material to generate holding power. They should not be used in concrete or masonry that is in poor condition.

Plastic or Nylon Anchors. Highly economical, plastic anchors are used for mounting any fixture or hardware normally held by wood screws. They can be used to hang pictures or mirrors or mount kitchen and bath accessories,

drapery hardware, and many other items. As shown they can be used in both solid and hollow wall applications.

The anchor consists of a hollow sleeve that is inserted in a predrilled hole in the wall. The hole in the center of this plug tapers so that it is narrower at the tip than it is at the surface. A lip keeps the anchor from slipping through the hole and covers its raw edges. Make the hole the same diameter as the shank of the fastener. Then, insert the wood screw through the fixture and drive into the anchor. The screw expands the back end open so that the entire assembly is wedged tightly inside its hole. Plastic anchors are available in various lengths and diameters to accommodate various sizes of wood screws and to handle varying loads. As always, hole size is critical. The anchor must fit snugly and it is only as good as the surrounding wall material. Plastic anchors will not hold against much horizontal pull, especially in crumbly masonry.

Lead Screw Anchors. These fasteners are similar to plastic screw anchors, except that they are used to fasten medium-weight objects

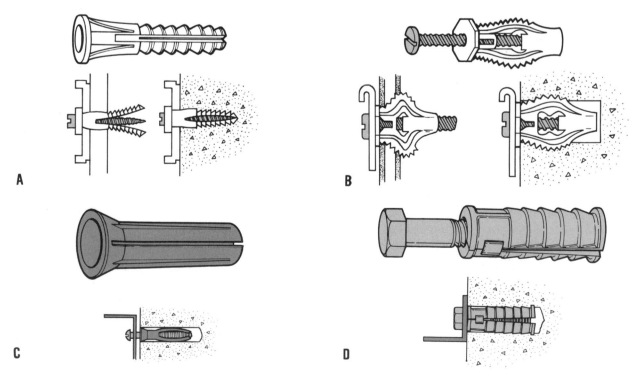

A   B   C   D

Various types of anchor bodies used with screw-in type fasteners: (A) plastic, (B) nylon, (C) lead, and (D) lag. Both plastic and nylon anchors can be used in solid and hollow applications.

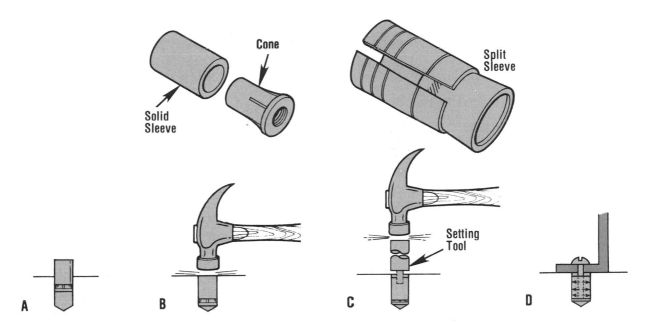

Steps in installing a caulking-type machine bolt anchor: (A) insert cone and sleeve in properly sized hole, (B) hammer sleeve flush to mounting surface, (C) use seating tool to countersink anchor below surface, or (D) position fixture and turn bolt or screw into anchor cone threads.

with wood screws. Lead anchor lengths commonly range from 3/4" to 2". Select a screw length that equals the thickness of the mounting hardware, plus the length of the anchor, plus 1/4". Drill the hole in the masonry surface the same diameter as the anchor and 1/4" deeper than the anchor. Set the lead anchor flush with the wall surface, and drive the screw through the mounting hole in the hardware and into the anchor with a screwdriver.

Lag Screw Anchors. These anchors, also known as expansion shields, consist of soft lead sleeves that can be attached with lag screws to concrete and other masonry surfaces. Specially designed horizontal fins keep the anchor from turning in the hole as the lag screw is tightened. Tapered annular rings also offer additional resistance to withdrawal from the hole. And because lag screw anchors are threaded internally to match the lag screw, they are quick and easy to install.

To install a lag screw anchor, drill a hole the same diameter as the outside diameter of the shield. Set the anchor flush with or slightly below the surface. Insert the screw through the fixture

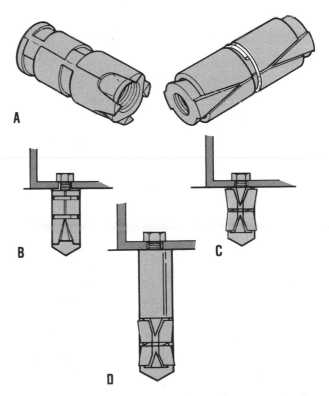

Noncaulking-type machine bolt anchors: (A) single and double designs, (B) single installation set flush, (C) double installation set flush, and (D) double anchor in deep recess using pipe sleeve

or hardware and into the anchor. Drive the lag screw into the anchor with a wrench. The anchor expands as the lag screw enters.

Machine Bolt Anchors

Machine bolt anchors are used to support objects that exert an extremely strong pulling force. There are two types of machine bolt anchors available: caulking and noncaulking. Caulking anchors are best if you are having difficulty drilling a hole of uniform diameter and true straightness. Noncaulking anchors are longer than caulking anchors, so they are a better choice when fastening to masonry of questionable strength.

Caulking Anchors. This anchor consists of a short lead sleeve that is slipped over a slightly tapered cap. The cap has internal threads to accept bolts. To install the anchor:

1. Drill a hole to the manufacturer's recommended diameter and depth. In weak or soft masonry or concrete, a slightly deeper hole may be drilled so the anchor can be countersunk below the surface.
2. Slip the sleeve onto the cone and insert the anchor cone end first into the hole.
3. Strike the sleeve of the fastener with several sharp blows of a hammer. This forces the sleeve down onto the wide end of the cone, expanding the sleeve so it presses firmly against the walls of the hole.
4. Special setting tools are needed to set the sleeve below the surface of the wall or floor.
5. Position the bracket or hardware over the anchor and install the machine bolt into the threads of the cone. The minimum length for the machine bolt is one that engages at least two-thirds of the threads in the cone after the hardware is in place.

Bolts can also be turned into the bottom of the cone prior to installing the anchor in the hole. In this way, a threaded nut can be used to fasten the fixture to the end of the bolt protruding above the surface. When bolts are installed prior to seating the anchor, the special seating tool must be used to prevent damage to the bolt threads.

Noncaulking Anchors. Noncaulking anchors are available in standard single and special double styles. These anchors are one-piece units, but their design is similar to the caulking anchors just described. A cone having internal threads is surrounded by an expansion shield or sleeve. As the machine bolt is tightened, the threaded cone is drawn up into the expansion shield, which develops a wedging action inside the precisely sized hole. These anchors develop holding power deep in masonry where the masonry material is the strongest.

Double-style noncaulking anchors are designed for use in concrete and masonry of questionable strength. They are also excellent when heavy shear loads are encountered. Setting the

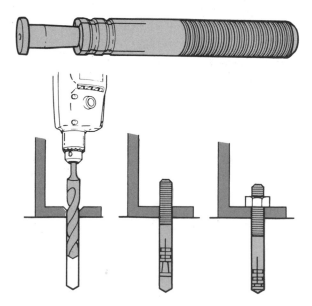

Typical one-piece stud anchor design and installation method

anchor forces the opposing wedges at either end to draw tightly into the anchor, resulting in full-length expansion against the sides of the hole.

To install either style of noncaulking anchors, proceed as follows:

1. Drill a hole to the manufacturer's recommended diameter and depth. Insert the anchor into the hole, making certain that the threaded cone is inserted first. For maximum expansion, the double style should protrude slightly.
2. Put the fixture or hardware in place, insert the bolt, and then tighten with a wrench until firm resistance to further turning is felt. The bolt should be securely tightened to develop maximum wedging action.

For maximum holding power, you may want to set a double-style anchor in an extra deep hole. When this method is used, it is recommended that a pipe sleeve of suitable size be set between the anchor and the fixture to be anchored.

One-Piece Anchors

There are a wide variety of one-piece anchors available for attaching hardware and fixtures to solid concrete and masonry.

Stud Anchors. Stud anchors resemble headless threaded bolts. When installed, a portion of the threaded anchor protrudes above the surface. The hardware to be attached is slipped over the anchor and secured with nuts and washers.

Like all solid-surface fasteners, stud anchors expand when set to press tightly against the sides of the drilled holes. Hole diameter and depth must match manufacturer's recommendations; there is little room for error with one-piece anchors.

Expanding or setting stud anchors may involve striking them with a hammer and/or

tightening down the fastener's nut. Matching nuts and washers may or may not be provided with the stud anchor. A tapered tip on the end of the anchor prevents damage to the bolt threads when the anchor is struck with the hammer.

Expansion Bolts. One-piece expansion bolts are made from a very tough, high-grade steel that is previously expanded during the manufacture of the bolts. When driven into the proper diameter hole in hard masonry, the two sheared and previously expanded halves of the bolt shank are compressed by the wall of the hole. Because they will forever attempt to re-expand to their original bulged shape, a tremendous holding force is generated.

These fasteners are meant for use in very hard materials, such as high-quality concrete, extra-hard brick, or stone. Do not locate them in mortar joints. As shown, round, countersunk, tie wire, and stud heads are available. Resistance to withdrawal is so great that the stud and nut design should be used if there is a chance that the fixture or hardware will be removed in the future. The only method of "removing" a one-piece expansion joint is to drill the hole extra deep during installation; the anchor can then be driven beneath the surface and patched over, if needed. Mounting holes should always be deeper than the length of the anchor. To work proper-

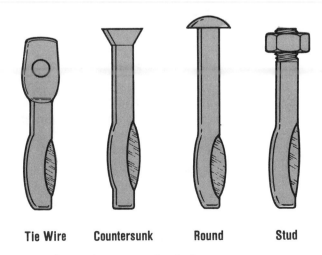

| Tie Wire | Countersunk | Round | Stud |

Typical one-piece expansion bolts

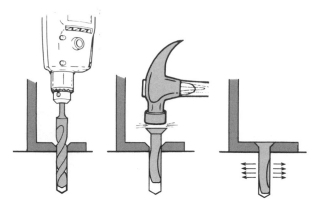

Expansion bolt installation

ly, one-piece expansion bolts must not touch the bottom of the drilled hole.

Nail and Riveted Anchors

For light work in solid and hollow walls, nail-in fasteners with a nylon anchor body are ideal. A hole is drilled the same diameter as the nylon anchor, which is then inserted into the wall. The partially threaded nail is then driven into the anchor body to expand it against the sides of the hole. If the fixture must be removed at a later date, the threaded nail can be turned out using a screwdriver and the nylon anchor easily pulled from the wall.

Another type of nail-drive fastener resembles an oversized rivet and works on the same principle.

HOLLOW WALL ANCHORS

As shown earlier, a number of expandable anchors used in solid concrete and masonry can also be adapted for use in hollow walls. In addition, two major types of fasteners are used exclusively in hollow walls: toggle bolts and expansion anchors.

All hollow wall fasteners rest in a hole drilled completely through the wall. Wings or legs on the end of the fastener are then set or expanded to tightly grasp the wall from the backside.

The thickness of the wall will determine the length of the fastener used. An easy way to mea-

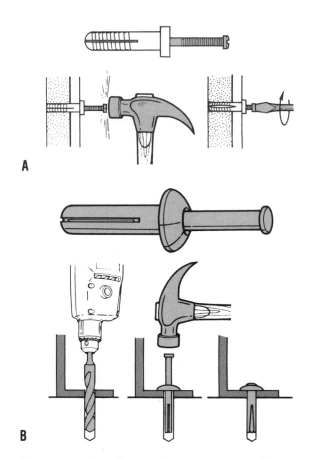

Two types of nail-in anchors: (A) removable screw head, and (B) permanent rivet head

sure wall thickness is to insert a crochet hook or hooked wire through a hole drilled in an inconspicuous area. When the hook catches on the back face of the wall, mark the wire at the hole opening. Withdraw the wire or crochet hook and measure the distance between these two points.

When working in hollow core block, locate mounting holes away from the block's solid webs or any grouted cavities. If the hole is drilled too close to a web, the web may interfere with the seating of the anchoring wings or legs.

Toggle Bolts

Toggle bolts use spring-loaded wings threaded onto various types of mounting bolts. The wings are folded back against the bolt that is then inserted through the wall hole. When the wings emerge from the backside of the drilled

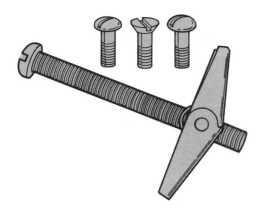

Typical toggle bolt design with removable wings

hole, they spring open to create a bearing surface against the backside of the wall.

Because they spread the load over the entire length of the wings, toggle bolts are especially suited for attaching objects to soft, crumbly material. They can be used to provide good anchoring in concrete block, thin cavity walls, hollow tile, stucco over wooden sheath, and drywall construction.

To install a typical toggle bolt, drill an oversized hole large enough to admit the wings when folded. Then proceed as follows:

1. Turn the wings off the bolt portion of the fastener, and insert the bolt through the mounting hole in the item to be secured to the wall.

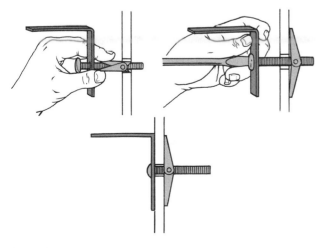

Toggle bolt installation

2. Screw the wings back onto the end of the toggle bolt, and fold them back completely.
3. Lift the item to be mounted into position, and insert the toggle bolt into the mounting hole until the wings spring open on the backside of the wall.
4. Pull back slightly on the assembly to hold the wings against the wall and to prevent them from turning when the bolt is tightened down.
5. Turn the bolt with a screwdriver or wrench until the mounted object is firmly set against the wall. When the installation is complete, the wings of a standard toggle bolt form a perfect 90° angle with the inside of the wall surface.

Although they provide great holding power, toggle bolts are not without disadvantages. If several toggles are needed to secure a large item, all wingless bolts must be inserted through the mounting holes in the item at the same time. You may also need more than one set of hands to hold the item in position as the wings are folded, slipped into the wall holes, and tightened down. Extreme care must also be taken when locating wall holes for multiple toggle bolt installations. The most accurate method is to hold the item or fixture in final position and mark all hole locations at one time.

A second disadvantage is that if the toggle bolt is ever removed from the wall, the wings slip off the end of the bolt and are lost in the wall cavity. Obviously, toggle bolts are best suited for permanent installations.

Expansion Anchors

These screw anchors, also known as wall expansion anchors or Molly-bolts, are extremely popular and easy to use. When the screw of the anchor is turned home, the legs of the fastener expand and are drawn up tight against the inside wall surface of the hollow material.

The anchor is one piece, and there is no need to compress the leg to fit it through the mounting

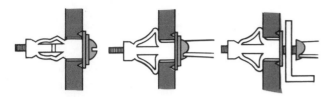

Hollow wall expansion anchor installation

Attaching a steel plate anchor with adhesive

holes. Unlike toggle bolts, once a screw anchor is set, the screw may be removed as often as you like without affecting the holding power of the anchor portion.

Expandable leg anchors must be correctly sized to wall thickness. The smooth section of the anchor shank should equal the thickness of the wall. Anchors that are too short or too long will not generate enough holding power.

Hole diameter must also be precisely sized to match anchor size. Once the hole is drilled, insert the anchor into the wall. If the anchor has pointed prongs to help it grip the outside wall, tap the anchor with a hammer to embed the prongs in the wall. The heel behind the screw head should be flush with the wall.

Turn in the screw until you feel a definite resistance, which indicates the legs have expanded and the anchor is set. You can now remove the screw from the anchor, position the item to be mounted, and reinstall the screw.

FASTENING WITH ADHESIVES

Adhesive technology has evolved to the point where high-strength glues and mastics now offer a viable alternative to certain mechanical fastening techniques.

Flat, perforated steel plates with nails or bolts projecting from their faces can be attached to concrete and masonry using special mastics. Once the mastic has set, lumber can be impaled onto the nails or hardware that is secured to the bolts with nuts and washers.

Modern construction and panel and foam adhesives can also be used to attach wooden furring strips or rigid foam insulation directly to concrete, brick, or block. Always follow manufacturer's suggested applications, as solvent-based adhesives will destroy many types of foam insulation board. Wood paneling should always be fastened to furring strips when installed over a concrete or masonry wall. Never attempt to glue paneling directly to concrete or masonry.

Before using adhesives on concrete or masonry, be sure the surface to which you will be bonding is structurally sound, clean, and dry, and is free of grease and loose paint. With poured concrete foundations, it is especially important that the surface is free of any form release agents that may have been used during construction. These agents usually contain oils, greases, and/or silicons that can adversely affect bonding strength. To remove them, wash the surfaces thoroughly with a trisodium phosphate (TSP) solution or strong detergent. See chapters 10 and 11 for more details on cleaning and repairing damaged concrete surfaces.

GLOSSARY

Adhesion. The sticking together of substances in contact with one another.

Admixtures. Any materials, other than portland cement, water, and aggregates, that are added to concrete, mortar, or grout immediately before or during mixing.

Aggregate. Bulk materials, such as sand, gravel, crushed stone, slag, and pumice that are used in making concrete.

Anchor Bolts. Any of a variety of rather large "J"- or "L"-shaped bolts designed to be partially embedded in concrete or mortar.

Ashlar. Quarried building stone that is squared on all edges.

Backfilling. The process of piling earth against the outer surface of a form.

Base Coat. Each of the lower layers of plaster, if more than one coat is applied.

Batter Boards. A board frame supported by stakes set back from the corners of a structure to allow for relocating certain points after excavation. Saw kerfs in the boards indicate the location of the edges of the footings and the structure being built.

Bed Joint. The horizontal layer of mortar on which a masonry unit is laid.

Block. A concrete masonry unit made with fine aggregate and cement; any of a variety of shaped lightweight or standard-weight masonry units.

Bond. The property of hardened mortar that holds the masonry units together. Also, the lapping of brick in a wall to provide solidity.

Brick Masonry. A type of construction consisting of units of baked clay or shale of uniform size, small enough to be placed with one hand, and laid in courses with mortar joints to form walls of virtually unlimited length and height.

Brick Set. A wide-blade chisel used for cutting bricks and concrete blocks.

Brown Coat. The second coat of plaster or stucco in three-coat work.

Buttered. The small end of a brick that has a quantity of mortar placed onto it; the act of buttering a brick's end with mortar.

Caulk. To seal up crevices with flexible material so as to prevent moisture entry.

Closure Brick. A partial brick that is cut to fit in order to complete a course.

Coloring Agents. Colored aggregates or mineral oxides ground finer than cement and added to concrete.

Concrete. An artificial stone made by mixing cement and sand with gravel, broken stone, or other aggregate. These materials must be mixed with sufficient water to cause the cement to set and bind the entire mass.

Control Joints. Continuous joints built into concrete slabs or concrete block walls used to control or prevent cracking. Wall joints permit slight movement without cracking. Slab joints force the crack to occur at the joint location.

Corbelling. Courses of brick set out beyond the face of a wall in order to form a self-supporting projection.

Courses. One of the continuous horizontal layers (rows) of masonry that form a masonry structure.

Curing. The process of protecting concrete against loss of moisture during the early stages of setting; usually accomplished by covering it or sprinkling it periodically with water.

Dry Mixture. A mixture of concrete whose water content is severely restricted.

Earth Forms. Used in subsurface construction where the soil is stable enough to hold the concrete while retaining the desired structural shape.

Edger. A concrete finishing tool for rounding and smoothing edges, which serves to strengthen them.

Edging. The process of rounding the edge of freshly poured concrete; one of several finishing techniques.

Efflorescence. A powdery stain, usually white, on the surface of or between masonry units. It is caused by the leaching of soluble salts to the surface.

Expansion Joint. A material placed within or a scoring of concrete that allows it to expand without cracking.

Exposed Aggregate. A concrete finish achieved by embedding aggregate into the surface, allowing the concrete to set up somewhat, then hosing down and brushing away the concrete covering the top portion of the aggregate.

Face Brick. A type of brick veneer made specifically for covering walls.

Fieldstone. Common stone found in fields and along rivers and streams; used in masonry.

Finish Coat. The top layer of plaster if the plaster is applied in more than one coat.

Flashing. A waterproof covering placed at certain points in brick masonry to hold back water or to direct any moisture outside of the wall.

Float. A wooden tool used to finish a concrete surface.

Footing. A base for a wall or other structure that provides stability.

Form. A parameter or set of parameters made from earth or wood to hold the footing concrete.

Frost Line. The maximum depth to which frost normally penetrates the soil during the winter; varies from area to area depending on the climate.

Furrowing. Striking a "V"-shaped trough in a bed of mortar.

Gradation. The distribution of particle sizes, from coarse to fine, in a given sample of aggregate.

Grout. A water–cement, or water–cement–sand mixture, used to plug holes or cracks in concrete, seal joints, and fill spaces.

Hawk. A fairly small board with a handle that is used for holding mortar.

Header. A masonry unit laid flat with its longest dimensions perpendicular to the face of the wall; generally used to tie two wythes of masonry together.

Hydration. The chemical reaction that occurs when water is added to cement, causing it to harden.

Joint. Any place where two or more edges or surfaces come together.

Jointer. A tool used for making grooves or control joints in concrete surfaces to control cracking.

Lintel. A beam placed over an opening in a wall.

Masonry. A type of construction made of prefabricated or natural masonry units laid in various ways and joined together with mortar.

Moisture Content. The amount of water contained within the aggregate used in concrete.

Mortar. A mixture of cement, sand, and water without coarse aggregate; used chiefly for bonding masonry units together.

Parging. Covering a masonry surface with a thin layer or layers of mortar to seal it against the effects of moisture.

Pavers. Bricks of numerous size and shape that are used in constructing sidewalks, patios, and driveways.

Pier. A free-standing column.

Pilaster. A projection from a masonry wall that provides strength.

Plumb. That which is vertically perpendicular as measured with a spirit level or plumb bob.

Pointing. The process of inserting mortar into horizontal and vertical joints after a masonry unit has been laid.

Portland Cement. A number of cement types with unique characteristics manufactured from limestone and mixed with shale, clay, or marl.

Precast Concrete. Concrete that is cast in forms at a place other than its final position of use.

Premix. Packaged mixture of ingredients used for preparing concrete or mortar.

Reinforcing Bar. A steel bar that is used for reinforcing concrete and masonry structures.

Retaining Wall. A wall that is constructed to hold soil in place.

Rowlock. A brick laid on its edge (face).

Running Bond. A simple bricklaying pattern featuring continuous horizontal joints and vertical joints which are offset or in line.

Scratch Coat. The first coat of plaster or stucco.

Screed. A long, straight board used for striking off concrete.

Screeding. The process of leveling the surface of a concrete slab by striking off the excess concrete.

Set. The process during which mortar or concrete hardens. Initial set is present when the concrete has to be broken to change its shape, generally about an hour after it is placed; final set occurs about ten hours after placing the concrete.

Shell. The sides and recessed ends of a concrete block.

Soldier. A brick laid on its end so that its longest dimension is parallel to the vertical axis of the face of the wall.

Stretcher. A masonry unit laid flat with its longest dimension parallel to the face of the wall.

Striking Off. The process of removing excess concrete to a required level.

Stucco. A finish composed of two or more layers of mortar that is applied to either indoor or outdoor walls.

Tamp. The process of compacting concrete with rakes or short lengths of lumber.

Texturizing. Creating a particular finish in a concrete surface, such as brushed, smoothed, etched, or pockmarked.

Ties. Wires or rods that are used to hold masonry structures together at specific locations.

Trowel. A steel tool with a flat surface that is used to smooth concrete.

Tuckpointing. The process of refilling old joints with new mortar.

Veneer. A layer of bricks or stones that serves as a facing.

Wales. Horizontal members that aid in wall/form reinforcement and distribution of forces.

Weep Holes. Openings made in mortar joints to facilitate drainage of built-up moisture.

Wire Mesh. Any of a variety of types of bonded wire forming a mat used to reinforce slabs of concrete.

Workability. The ease or difficulty of placing and consolidating concrete.

Wythe. A vertical stack of bricks one thickness wide.

INDEX